Still Colored

COMFORT GREEN

PAGE PUBLISHING
Conneaut Lake, PA

First originally published by Page Publishing 2023

ISBN 979-8-88654-797-9 (pbk)
ISBN 979-8-88654-798-6 (digital)

Printed in the United States of America

Acknowledgment

This story began over forty years ago, but I did not start writing until about eighteen months ago, and it basically depicted my travels and experience throughout most of the world. A lot of people have the same experience I went through and surely dealt with their situations in a different way or maybe the same way with different outcomes.

I would like to dedicate this book to my parents, Thomas Sona Ngu and Esther Mafor Ngu, who are the greatest parents anyone can ask for. My parents taught me to accept Christ as my Lord and Savior as well as believe in myself. They let me believe that anything is possible if you work hard and believe in yourself, have a good character, and demand what is right.

I would also like to thank my children, Amie, Spurgeon, and Thomas, for being the best children anyone can ask for; my siblings Mary, Martin, Samuel, Robert, Paul, Elizabeth, Joseph, Emmanuel, Rosalyn, and Dr. Lawrence Ngu; and my countless nieces and nephews. Spurgeon Green III for giving me three beautiful children. My sister Margaret Thomas Ngu for her unconditional love and support.

Thank you to my lawyer, Attorney Cristina Carabetta, for being there for me for legal and nonlegal issues. A special thank you for the countless people who have crossed my way during my journey. The many friends I came in contact with throughout my journey: Mary Anne Shepherd, an angel from God; Willie Faust, my biggest fan; Dr. Stephanie Brown. To my friend Mrs Jane Yongo for being my eyes and ears and for her devoted support. My friend Sabina Amporful a sister and confidant. My friend and doctor Flemming Burroughs and his wife Dr. Rosie. To my friend and doctor John D. Marshall who died from covid. You were my guiding light and this book is also dedicated to you my hero. And countless of friends that supported and worked with me throughout this journey. Thank you, Dr. Evaritus Oshiokpekhai, DMP, and wife for taking care of my feet so I can wear high-heeled shoes. Dr. Audrey Hodge, MD, my friend who trusted and supported my leadership during our time at Sumter Regional Hospital. Thank you to Dr. H. Katner for being my friend and support system, supporting and working with me throughout this journey. And last but not the least, to Dr. Spurgeon Green Jr., my father-in-law. A man of integrity and patience who devoted his life literally for the sake of his patients and treated hundreds and thousands of them without their ability to pay. In my humble opinion, very few people have done for humanity what he was able to do. He is an unsung hero.

I thought about writing a book a few years after I left SRH (Sumter Regional Hospital) about the struggles I had with understanding why virtual segregation is still so prevalent in our society.

My story begins in Cameroon, West Africa, where I was born. I left the country at an early age for Europe to study because my parents felt it was a way out of poverty and a way to make a future not only for myself but for my entire family. The thought of leaving Africa was one that was mentioned but never assumed would happen. In my case, it was possible because my father was an educator who had work with the Presbyterian mission as a school teacher, principal, and superintendent. Also, growing up, he made us believe that everything was possible if you believe in yourself regardless of your gender. As the years went by, I was surprised when one day, my father asked me how I would like to go to England to be with one of my siblings and work as a babysitter. This, in fact, was the only way possible for me to leave the country. As a young child, I was not sure how it would happen. I liked the idea but could not see how possible it would be because we did not have the kind of money to send me out of the county, and besides, I had older siblings who, by tradition, should have left before me.

Being from a Christian family, I hoped and waited for the day this miracle would happen and did not know it would be a reality. One evening in the early seventies, my older sister came to my boarding school for a visit and out of the blue told me I was leaving for England. I did not fully understand how it was possible, or if she had lost her mind because I knew we did not have the finances of fully taking care of the large family we had and talking about sending me out of Africa. I had never traveled outside more than two hundred miles of the city I grew up in, so going abroad was a dream that I could not imagine.

As I prepared for the journey, I started feeling the anxiety and fright to leave the only environment I had known and started wondering if it was a good idea for a young girl to leave the only environment she was used to. My mother, who was trained by Swiss missionaries, told me going abroad will give me the life I needed and one that only a few will ever experience. At that point, just knowing I was going to travel gave me the sense of achievement and put me in a category as big as Oprah. As the news spread in our community, I became s celebrity overnight. Everyone wanted to see this little girl that was going to get in a plane and fly to the "White Man Country." I became an instant celebrity. As the day for my departure got closer, I was taken out of my boarding school and got the necessary documents for me to leave for England.

I arrived in England as a frightened little girl after flying for the first time and was met by a family I did not know. These people had not seen me either, but carried a sign with my name to let me know they were there to take me to my final destination. As I got off the plane, I had never seen so many White people in my life. As a child, I was led to believe that White people were next to God, and when I saw a lot of them at one time, I felt I was on my way to heaven. I could not understand what was happening, but I knew my struggles were over and my life was just beginning. The unknown family picked me up, and unfortunately, I could not understand what they were saying although we both spoke the English language. We decided to write out what we were saying to each other. I had a bag full of paperwork giving me instructions on what to do when I arrived my destination. I gave the information to my host family, and they took me to their house in East London.

As I got to their house, I was given a room to sleep, and for the first time, I slept on a mattress that was so soft, and once again, believed that I was going to heaven. I had never slept in a room by myself and could not imagine an environment so clean and beautiful. The lawn was well-manicured, there was running water in the house, they had a bathroom and toilet in the house, and no children but the two adults. I say this because it was common to have an entire family sleep in the same room or even on the same bed or for that matter, sleep on the floor. Sanitation was not something that was considered normal or enforced in the entire country. It was common to see men and women use the grounds to urinate, likewise with children. You could smell the odor everywhere you went and accepted it as part of life. I slept all night after an eight-hour flight, and when I woke up, I was given breakfast (toast, eggs, and tea), which I was not used to because at home we only had tea for breakfast. After eating, my host family told me they had contacted my brother who lived about seventy miles away, and he was on his way to pick me up. I had not seen my brother for over ten years, and was happy to see someone who looked like me and could understand me.

The first few months was very difficult for me because I missed my parents and could not understand why there were so many White people and most of them were doing work that was supposed to be done by Blacks. Growing up, I was told that White people did not wash dishes, cut grass, sweep floors etc., so it was confusing for me to see so many doing that exact thing I was told they could not do or should not do. In my community, when I was growing up, you could count the number of White people you saw in one hand. They were either there as missionaries or owners of Christian schools. They were the only ones that owned cars and lived in houses that had running water and lots of servants. My family was considered privileged to be in the company of Whites because my father and mother were trained by them and assisted the Whites with the culture. They were the go-between for the tribal chiefs and the politicians.

As the days and months went by, I started understanding that I was in the minority and started questioning why my parents sent me out of the country. I knew it was for the best, so I tried to understand and knew that it was better than being in Africa. After a few months, I registered in a high school to take secretarial courses because I was told the only job I would be good at was a secretary. I took those classes, and within six months, I received a diploma and got a job as a secretary for an old White man in a cement company. My duties were taking notes in shorthand and making tea for the entire office in the morning and in the afternoon consistently every day. After six months with the company, my brother left for the United States of America and told me I could stay in England or go back to Africa. I could not imagine going back to Africa, so I informed him I will stay and make the best out of the situation. My father had instilled in me that going abroad (White Man Country) was an opportunity of a lifetime. It was an avenue for me to better not only myself but put myself in a position to help my immediate and extended family members.

As the months went by, I had to rent a flat and pay my bills with a very minimal income. I found a flat that did not have central heating but was affordable. As the months went by, I was getting used to the situation and had the same routine of going to work and school. I kept my parents informed of what I was experiencing and how I was homesick but knew I had to make it work. I knew there was no turning back. I found myself depending on a lot of people for basic needs; for example, showing me how to catch a bus to work and school and also helping me with finding medical help if needed.

As months and years went by, I could not handle staying in England due to the fact that I did not have the necessary documents to live there without attending college. The prime minister passed a law that foreign immigrants without the proper documentation had to leave the country. Once again, I had to turn to my brothers in the United States to find a way out for me. I was determined not to go back to Africa. I had gotten used to the finer things in life—for example, riding on a bus, running water (which we had at home but not as plentiful), clean environment, and actually attending a university. This was not a life I wanted to let go. After almost three years of being on my own, I started developing a strong will and a fighting attitude where I told myself I was born to be great and had to stay on track. I decided to visit the US embassy in London and find out about visiting the US as a student. I also checked with my brothers in the US and had them secure a community college and a host family that would sponsor my stay in the Unites States of America.

After three months of working on trying to get out of England, I received a letter of deportation from the British Government saying

I was going to be deported if I did not have a full-time student status. At this juncture, I could not afford tuition for a full-time student and my only recourse was to try to get out of the country before deportation. Within three months of my letter of deportation, I received a form 120 from a community college in Worthington, Minnesota, and a letter of admission to their institution. The next hurdle was to secure a ticket to fly out of England. Fortunately, my brothers came to the rescue with a ticket for me to leave for the US. I said my goodbyes to a few friends I had met and thanks to a special lady who became a surrogate mother to me. I had just one suitcase with all my belongings which had been accumulated for three years, but considered myself lucky and heading to the next chapter of my journey.

As I arrived the US as an adult, I was determined to start another chapter of my life and thank God for sparing me from deportation The United States was truly the land of opportunity, and if you eventually make it here, you have made it big. Upon arrival, I took a Greyhound bus from Chicago to Worthington, Minnesota. This time, I knew my host family and was more mature and used to seeing and being with White people. As a matter of fact, I had seen very few nationalities in the past three years. I made it to Minnesota in September of 1976 and was happy to be welcomed with open arms. My host family were elders in the Presbyterian Church and had a successful photography business. They immediately got me into the school, and as an added benefit, got me involved in the church. I joined the church choir and felt at home because as a child, my parents made all of us sing in the church choir.

The next point on the agenda was to register for classes at the community college. I quickly made friends with the professors and was a sore thumb because I was the only Black student in the entire college campus. As usual, I was approached to play basketball for the college. I had never played basketball in my life, but this was an opportunity for me to fit in. The coach approached me and assumed I could play just because I was Black. I tried out for the team and could barely bounce the ball, but they just knew I could play. After a few months on the team and no play time, the coach still kept me on the team. This was my opportunity to get seen and be accepted.

I was approached by so many people just because I was on the team and excelled academically. That same year, our team won the State Championship, and as a team member, I became almost famous and accepted.

I was able to clef out of a few classes and graduate with an associate degree after one year at Worthington Community College.

The next journey was to find another school to attend to meet the requirement of an I20 visa. The obvious transfer was to Mankato State University in Mankato, Minnesota, because most of my basketball teammates were going there, and as a basketball player from a state championship team, I was going to be an easy recruit. In the back of my mind, I knew I needed to make the move but did not know who I was going to pay for it. My parents obviously did not have the funds to sponsor me, and I did not qualify for in-state tuition nor have the ability to work in the US. Once again, my faith had to take the upper role. It was another milestone to cross, and I was determined to find a way. My plans were to get admitted then worry about the tuition later. I applied for admission to Mankato State and was accepted for the fall of 1977. This was the first opportunity to go to a university and start the journey of becoming a lawyer, as promised to my father, or get a master's degree. I had three months to come up with the tuition of $900 and another $1,500 for room and board. All I needed was my tuition in order to satisfy my immigration status.

By the grace of God, when school started, I walked into campus the first day and checked into my dorms and found out that I had been admitted as an in-state student and only had to pay $205 for tuition, and make arrangements for my board and meals within the first four weeks of classes. I had chosen a room with three other girls, which would reduce the tuition, and no meal plans because I could not afford room and board. I called my host parents from my community college and informed them I was at school, and they offered to pay for my tuition for the semester. Once again, I dodged the bullet, and God was watching out for me. It was my turn to make sure I took the right classes and start my politic with the professors. The game plan was to visit my professors and let them know I was ready and willing to learn and contribute to the college in any way possible. I looked around and noticed there were very few minority students, and the focus was on me. I had a lot of people asking me about my name (Comfort) and wanting to know why I came to as little a town as Mankato in Minnesota instead of going to Minneapolis. This was the perfect plan for me because the attention was all mine, and I could use it to my advantage.

I had to think ahead of the next plan on how to get my tuition for the next fall quarter session. I knew my parents could not help me financially so I turned to the church. It was very easy for me to contact the church, being a Presbyterian, because there were no Black Presbyterians in Mankato and it was unusual to see one.

I immediately went to the pastor of the local church and became a member. They were excited to have a foreign student in town and wanted to help. I told them I wanted to help with the Sunday school kids and join the choir. This was an avenue for me to find work and secure funds for the next semester. The focus was to think ahead since I did not have a permit to work and could not get help from my parents back in Africa. The church gave me a job to type the bulletin for the Sunday Service and paid me $50 a week. They also arranged for me to babysit for the children of several members of the church. The future was looking bright because as long as I could make a little money for college, I was assured of staying in school and in the USA.

Life progressed for a while with minimal jobs until I met the founder of a large company in North Mankato, Minnesota, who asked me to work in his factory. I knew I did not have the correct documentations, but took the job. I did not have a means of transportation to and from work so I had to walk from Mankato to North Mankato every day for work. The job was in their factory, printing wedding invitations paying $44.35 per hour. This was the break I had been looking for, but I still had the problem of transportation. You see, Mankato did not have public transportation, so I had to walk or catch a ride to and from work. Needless to say, I had no one to help me, so I walked about 10 miles one way each day to work after classes. During the winter months, I had to walk lase in the afternoon and did not get to my dorm room until after dark. I asked

God for his protection and continued to think positively. This situation began my love for walking which I have kept thirty years later.

I struggled in college for the next two years given the fact that I did not have the time to study, therefore, my grades were not up to par. Upon graduation, I moved to Minneapolis where I stayed for one year waiting for my next direction from God. After not checking in with the university, I received a letter that if I did not apply for graduate school, I had to leave the country. I went back to Mankato and met with my professors and explained what was happening and begged to get back in school. I was given an unconditional admission and asked to score a 3.5 grade point average to be accepted into the Masters of Business Administration Program. I accepted the offer and starting searching for a place to live as well as money to pay for my education. I had a friend from college whom I called and explained to her what was going on. She offered for me to come stay with her for a month. The problem was she lived 14 miles away from campus and I had no means of getting to school. I accepted her offer and prayed for the Lord to come up with my next move.

The next few days, while I was waiting in Good Thunder, a suburb of Mankato, I happened to read the local newspaper and found an advertisement for a live-in sitter for the wife of the athletic director of the university who had just passed on. I immediately called and scheduled an interview to meet this eighty-two-year-old White lady named Rachael Wigley.

As I approached her house, I did not know what to expect but hoped for the best. When Rachael saw me she wanted to know why I was looking for a job and why I wanted to stay with an older lady. I informed her I was a foreign student from Africa and could not work in the country so I had no choice but to find anyone who will let me stay for room and board. I informed her I was from a Christian family and asked her to please help me get an education so I can help others. She immediately hired me and gave me a home for the next four years.

After graduation with a master's degree in business administration, I still had not acquired the necessary documents to stay in the country. The only other option was to become a professional student. To my amazement, my sister had a surprise for me. She had arranged for my parents to visit the United States to celebrate my graduation from graduate school. I made the trip to Aurora, Illinois, to meet with the family and see my parents after leaving home more than a decade ago. It was at this time that I was introduced to a Dr. Spurgeon Green Jr., MD and his family who had been taking care of my family free of charge. Dr. Green was very impressed that I had obtained a degree in business administration and wanted to know if I would move to Joliet, Illinois, and work for him. He indicated he was in the process of starting the first Black-owned HMO. Dr. Green graciously invited my family to join his family for dinner in a Chinese restaurant in Chicago. At this dinner, he introduced us to his only son, who was attending medical school in the Dominican Republic. Both families had a good dinner and exchanged stories.

When we came back to my sister's house, my parents asked me about the very handsome young man we met. At this point in my life, I was not looking for a husband or a relationship. I was focusing on a job and looking forward to earning an income which would enable me to help the family back home.

After a few days with my parents, I left and returned to Minnesota to continue my search for employment or continue to take more courses at the university. I continued to live with Rachael and started another program in political science in order to keep my immigration status current. This, in fact, was the best scenario for me because I had a comfortable home to live in as well as take care of a woman I called my mother. I had fallen in love with her, and although she never had children of her own, she treatment me as her child. As Rachael aged, I felt trapped about my next move. I did not want to leave her by herself, but she insisted I move on. I decided to find another student to take over my responsibilities of taking care of Rachael in the event I had to leave. As soon as I made arrangements with one of my friends from Taiwan and became a resident of the United States, I informed Rachael that I wanted to find a job and move closer to Chicago where my family was. She agreed with me, and I started looking for employment. With my MBA in hand,

I applied for a job and got a position as a claims adjuster with State Farm Insurance Company in Madison, Wisconsin.

The day of my departure was one of the toughest days I had in my life. Rachael and I cried, but she insisted it was time for me to move on. She also advised me to start thinking of finding a mate, which was a good idea because I had not dated anyone seriously. I could not afford to have a relationship with anyone because I did not have the time and was not sure if I could be accepted by the White race since they were the only race I had come to identity myself with.

I arrived Madison, Wisconsin, ready to start a new chapter in my life. The city was much bigger than Mankato, Minnesota, and my future looked bright. I had secured an apartment five minutes away from the office and was eager to start working and earning my own money. I spent the first few weeks at work getting to know my coworkers and studying the processes of being a claims adjuster. I realized that most of the adjusters at this particular office were lawyers or had advance degrees which made me feel challenged and obviously impressed.

As time went by, I was informed that I was on probation and had to attend three weeks of schooling in Bloomington, Illinois, and receive a certification for claim adjusting. I went to the training and received my certification and became a permanent employee. While I was still working with State Farm in Madison, I received a call from Dr. Green's son who informed me his father had opened his HMO and wanted to know if I was still out looking for a job. He also wanted to let me know that he was going to be in the area and would like to take me out for dinner. I informed him of my recent employment and accepted his invitation to go out. We went to Wisconsin Dells and had lunch, and another invitation was extended to me to go to a play in Chicago. I was very excited because he seemed very nice and educated and came from an upper-class family. My thoughts were along the lines of finding a mate that was self-sufficient and interested in the same things I was. I found out that he was a good Christian and was looking for a woman that worked as hard as his mother. After nine months of dating, we got married in Aurora, at the Presbyterian Church where my sister and her family worshiped. I submitted my resignation to State Farm Insurance Company and moved to Joliet, Illinois, with my new husband to start the next chapter in my life.

We moved in with my husband's family, and I took a position as a marketing director at my father-in-law's clinic. I was given a huge task immediately to managed his HMO as well as his medical clinic. This position was new to me because I had to learn about the health industry. It was a challenge for the first few months, but I had an excellent teacher. I was responsible for securing contracts for the HMO as well as recruiting physicians to work at the clinic.

My husband decided to attend law school in Chicago, which was about forty miles from Joliet. I did not get to see him as much since he attended night classes. For a while, it was difficult on me because I was home by myself most of the time and found myself wondering if I made the right decision to work for his parents, who were serious workaholics. I was fortunate that my father-in-law was a great teacher who took a lot of time to help and educate me under-stand the business.

We had our first baby, Amie Marie Green, in May of 1988. She was the most beautiful baby and the first grandchild to my in-laws. She was a joy to my husband and I and the perfect baby. Her father was so proud he basically spent every available minute with her. At three months, Amie was saying a few words already which was an amazement. In my opinion, she was the perfect baby. She seldom cried, and we did not have to child-proof the house for her safety.

We continued to enjoy Amie until her brother, Spurgeon Green IV, arrived seventeen months later. Spurgeon was delivered three months early and had a lot of medical problems. He stayed at a neo-natal hospital about thirty-five miles away from our home, which made it difficult for us to visit. My husband stayed by his side until he was deemed safe to come home. When Spurgeon Green IV was nine months old, my husband graduated from law school.

Three years later, we had our last son, Thomas-Sona Ngu Green. He was named after my father, Thomas Ngu Green. My husband decided to name our last child after my father because he wanted us to honor my father. With three children, it was time for us to talk about where we wanted to spend the rest of our lives. Our children were the focus of our lives, and we wanted them to have the best. We decided it would be better to move to the South, where my in-laws came from, to raise our children.

In 1995, we moved to Perry, Georgia, with our children and no jobs. At this point, our in-laws had already moved to Macon, Georgia, because they wanted to retire in the South and enjoy the warm weather. We felt comfortable because we had family in Georgia and the South was much safer in our opinion.

Neither my husband or I had a job when we arrived Perry, Georgia, but knew with our educational background, it would not take long to find a job. My husband started studying for the Georgia Bar Examination, and I started looking for a job. We found a three-bedroom house, which was considerably inexpensive compared to the houses in Joliet, Illinois, and made it our home. I spent lots of days sending out resumes and searching for whatever I could find to support the family before my husband passed the bar examination. We had a little savings, so we decided to take our time in looking for the right position. I wanted to continue working in health care, so it became my focus. I checked with my father-in-law, who was a local physician, for contacts in the area and his assistance in securing the perfect position. He asked me to look at the rural hospitals because he felt I had a lot to offer. My mother-in-law and I decided to take a trip to a rural hospital twenty-five miles south of Perry and inquire about a position as an office manager for a medical practice adver-tised in the local newspaper. She picked me up around eight o'clock in the morning and kind of instructed me that the South was a little different that the north. I did not fully understand why she made the statement at the point in time.

On our way to the location, I was wondering why the roads were narrow and all I could see was farm animals and fields of cotton and hay. I could not imagine what I was seeing. It almost occurred to me that I was back at home and wondered if I was still in the United States of America. I asked her why they had animals in the front of people's houses and why they were so many trailer parks. I also wanted to know if people lived in those houses and what they were used for. She did her best to give me a History 101 on the South and informed me that she grew up on a farm and that this area was predominately rural.

We arrived at the location for my interview, and I was more confused because the building looked like a barn and had a sign on it that said "Hospital." I said to myself, "What on earth is going on? Am I back in Africa?" I got nervous and starting wondering if we had made the right decision to move to Perry.

There was no turning back, so I had to let go of the negative thoughts in my mind and concentrate on why we were there. We walked into the hospital and asked to speak to the individual who had placed the advertisement in the paper. We were directed to a conference room where we waited for almost an hour before someone showed up to speak with me. I gave them my credentials and went through the necessary formalities of getting a job.

A few days later, I was called for a second interview and offered the position as office administrator of a hospital-owned practice. The salary was minimal but I was told it was comparable to salaries in that area.

The first day, I arrived at work all dressed up in a suit and ready to make a contribution to the world. I thought with an MBA, and many years of experience, I will not only have the say-so but will be approached for advice and recommendations on how to run the practice. Needless to say, it did not happen that way. My husband had instructed me, when asked where I was from, to say Mississippi and not to say Chicago. I did not understand why but told him with my accent, nobody will believe I was from Mississippi. He told me to fake it. He also said folks in the South are leery of people from the north. Being an aggressive person in nature, I did not listen to my husband's advice and ran my mouth when asked where I was from and told them Chicago. As I spent the first week at work, I was introduced to a few doctors and leaders of community who, in turn, asked me the very question of where I came from and told them again, Chicago. This old gentleman looked at me straight in the eyes and said, "So you are a damn Yankee," and then continued the conversation by saying, "How do you afford to drive a damn Mercedes?" It was obvious to relate the Mercedes to me because it had a plate registered in Illinois. I was, as usual, oblivious to what was said and remember that in all my life—even up to the point—I thought anything that was said by a White person was the truth and was considered a compliment. I could not imagine it being a derogatory comment, and still saw myself as superior and knew it all.

Even after being married to a family that was raised in the South and had gone through discrimination, I did not and refused to listen or understand their past experiences. I remembered my husband telling me that as a child growing up in Louisiana, with a father who was a captain in the Army and a mother with a degree from college, he still had to use a separate swimming pool for colored, which was the word that referred to Blacks when he was growing up. I am not sure if I just refused to believe what he was telling me or if it had been instilled in my mind that Blacks did not have to question what the White person said. On the other hand, I had not experienced any form of—what I felt was—discrimination. You see in Africa, the Whites were supposed to have the best foods, the best houses, best servants—virtually the best of all. The natives took all of their best crops to them and ate the crumbs that were left over. As a matter of fact, the Whites were superior in every respect. You were considered blessed to be in the company of a White person. It was a given, and there was no question or argument sort about treating or catering to a White person. It was instilled and reinforced in us that the Whites were superior, knew what was good for us as well as deserved the best. My point again was to chill and believe in what the White person told me instead of listening to my mother-in-law who had lived and experience the racial profiling during her entire life so far.

Each day went by, and I was faced with the second-class treatment but refused to accept it or understand. I continued trying to be the best employee, also remembering that my father told me I could be anything I wanted to be. My goal was to one day get a position as the administrator and felt it would happen. I knew I had the capabilities, and besides, the only people more qualified than myself were the doctors who only knew how to practice medicine but not run or manage an effective organization. Due to the fact that the majority (85–90) percent of the patients were Black, I could relate to them.

To my amazement, the Blacks questioned my abilities and wanted to know why I thought I knew more than they did and always wanted to know if I had checked with the doctors and gotten their opinion before relaying any answers to them. I could not understand, but as time went by, I realized that they might have taken the "colored" signs down, but the treatment was still the same. I also noticed that the Black patients were treated differently than the White patients. For example, the Black patients had to wait longer to see the doctor, they were seldom touched during their examination, and most of all, could not ask any questions about their treatment. I felt uncomfortable and started asking questions to the doctors about the time spent with the patients. I did not get a straight answer on why they were treating these patients like second-class citizens. It occurred to me that it was the way of life in South Georgia, especially in this clinic. I continued with my observation for a little while with total disgust and a sense of frustration.

The last straw was when I was told to cancel a Black patient's appointment and make room for a White patient because the doctor had to leave early and take care of his farm animal., and the White patient needed blood pressure medication. I informed the doctor the Black patient had to come in because of the severity of their condition and disagreed with his instructions to change the patient's appointment.

This bold expression of my feelings was not taken lightly by the doctor. The next day, I received a call from the administrator of the hospital and was given a verbal scolding never to question the doctor again. The administration did not want to hear the side of my story and told me, "It is different here, so do as you are told."

Being stubborn, I went back to the clinic and walked into the doctor's office and asked him what I did wrong. Because I was the administrator of the clinic and thought I was the boss, I did inform the doctor we had to do things the Christian way. This meant everyone was going to be treated equally regardless. I furthermore informed the doctor that we could not see his private patients on hospital time, and I had to reschedule his personal patients on the weekends. This did not sit well with the doctor. He took a lot of time off from the practice but wanted me to lie and say he was at the clinic. I did not feel comfortable with letting this doctor and his wife use the facility to run their personal business and be paid by the hospital when I knew it was unethical and wrong. The only option was to write a letter to the administrator and inform him of what was happening. To my surprise, I was immediately transferred to another doctor's office without notice and informed that the doctor did not want a woman to tell him what to do.

I started thinking of a way out of the situation because I did not like what I saw and was not going to be part of what was going on, which was letting a doctor take advantage of the system as well as be openly defiant and enjoyed beating the system. I felt I was doing the right thing by sticking with my guns, and after all, I had the MBA and felt everything was going to be okay. To my surprise, I was asked to move to another department without any explanation. I later found out from my successor that I was being too aggressive and did not know how to act. Basically what it meant was that I was not listening to the White doctor and was acting "sassy," which was unheard of from a Black person.

I came home and informed my husband of what had happened, and he once again told me "Welcome to the South." I still could not imagine or understand why I would be treated as such when I was performing my duties as asked and had just completed a favorable survey for the hospital which gave us a rural health status. I was instrumental in increasing the revenue for the clinic within nine months of my employment and reduced the turnover by almost 50 percent. With this incredible achievement in such a short time period, I thought I was doing a heck of a job. I wanted to think that being dependable and faithful would lead to respect, but to my surprise, I was not getting the result I expected. It took a long time for me to attribute the treatment I was getting to my race. The first thought that came to my mind was that I was a woman. I began to ask questions to some of the employees and got the same response, which was "listen to the White doctor." This comment was made mostly by the Black employees. You see, I did not see the racial division because everybody used the same bathrooms, waiting area, and ate on the same table during breakfast and lunch. Basically, we all shared the office, and there were no visual signs directed to any particular race. The last straw for me was when I was suddenly told to move to another department of the hospital and was not given proper notice. I was extremely hurt because I had bonded with a lot of my patients and knew they were not going to be treated properly. I left without a fight because I had to feed my family, and to be honest, could not understand what to do or who I could talk to about the situation. I had been told by my family that this was the South and to be prepared for this type of treatment.

I left the office with my belongings and was very unhappy but knew I still had a job. My husband was still waiting for his bar result so I had no choice but to continue working. I told myself Georgia is considered the Bible State so this situation I am going through must be an isolated one. In the back of my mind, I knew something was not right, but as a Christian, I could not bring myself to think I was being profiled. How could this situation be normal? After all, I was very educated, dressed appropriately, drove an expensive car, and my husband was an attorney. This was not what I had bargained for, and I was not going to tolerate the type of treatment I was getting.

I assumed the new position as the administrator of an orthopedic clinic and met with one of the most caring doctors. He was from the northeast of the United States and passionate about his patients. I became good friends with him and told him what had happened to me at the other clinic. He basically told me what I had been hearing for a few years now, but encouraged me to let it go. As the son of a preacher himself, he encouraged me to concentrate on the positive and not let the negative get the best of me. I took his advice and continued to find ways of improving his practice as well as revenue for the hospital.

As part of my responsibility for this new clinic, I was to attend continuing education classes in order to fully understand orthopedic specialty. A seminar came up at one of the most beautiful location (Cloister). I would be in one of the most beautiful resorts in the United States for a three-day seminar on coding. I submitted my request to the administrator and explained to him that I needed to attend this particular seminar. In my request to the administrator, I outlined the benefits of me attending the seminar and how it will help the clinic increase its revenue. Besides, it was a requirement that I take these courses to keep up with the requirements of my job. I received a response from my boss saying it was not necessary for to attend the seminar, and furthermore, it was too expensive, but the director of another department was going to attend the seminar and relay the information to me.

I marched to the administrator's office and told him I was not going to accept his justification of why I could not go to the seminar. I, for the first time, directly asked if it was a racial issue. This was a bold comment for me to make, but I was fed up with what was going on around me. I was beginning to see and experience what most people had been telling me of the racial divide, and that it was a battle I had to attack immediately, before it got any worse. I left my boss's office not knowing what he was going to do about my frankness, or better yet, my comments. At this point, I did not care because I knew I was telling the truth. I called my husband to tell him what had happened. He did not say too much about the situation but reiterated the fact that this is not uncommon in this area.

A few days later, I received my request signed to attend the seminar. This was a victory for me, and I immediately felt I had won the battle and things were going to change. I went to the seminar and learned a lot and came back ready to make changes and improvements to the practice. My first order of business was to visit the primary-care physicians in a fifty mile radius to get them to refer patients to our clinic. To accomplish this, I needed business cards and permission to travel outside of my location. To my surprise, my request was granted.

On one of my many trips, I met a pharmaceutical sales representative who informed me that he knew of a bigger hospital that was looking for a director of marketing for their hospital. He encouraged me to apply for the job and asked me to use him as a reference. I took his advice and immediately applied for the job. I waited for almost six months but did not hear from this hospital so I thought the position had been filled. Six months later, this same sales representative came to my office on one of his visits and gave me a newspaper article indicating this hospital was being criticized for not having any minority working in a management position. I realized it was the same situation with my current hospital but just attributed the fact that it was a very small hospital. As I thought about the situation, I realized that a lot of the non-revenue departments were staffed by Blacks. For example, environmental service department, Nutrition, and most of the Nursing. In the case of the nursing department, most of the aides were Blacks. When I posed the question why this was the case, I was told it was difficult to find qualified Blacks who wanted to move to rural areas and work. I had other battles to fight, so I decided to leave things alone and make the best of my situation.

A few months after, I received a call from my pharmaceutical sales representative about the public relations issue in the neighboring hospital. This time, the hospital was looking specifically for a minority to be part of their management team. He insisted I take a trip to the hospital and apply for a position there because the hospital was under tremendous pressure to hire a minority. I went to the hospital the next day and submitted my resume. Within a few days, I received a call to come for an interview for the position of marketing and public relations director. This is the same hospital that I had submitted a resume for a posted job one year earlier.

I went to the interview and was offered the job two days later. I accepted the job because it paid 30 percent more than what I was making, and it was a career advancement. I handed my resignation and thanked God I was not only moving to a bigger hospital but getting a better salary.

As I arrived at my new position, things looked bright. I had a story written about me in the local newspaper and immediately felt the warmness. As the days and months went by, I started realizing that some of the issues at the smaller hospital were the same at my present employment. I noticed most of the Blacks were saturated in the kitchen or were cleaning the hospital. I found out I was the only one in a management position and was the first minority hired in a management position in the history of the hospital. My immediate thought was to read more about why this hospital did not have any

Blacks in their management. I did not want to form an opinion, although I sensed it might be the same happening in the area. I kept an open mind gladly assumed my position.

In the weeks following, I received a welcome call from a local physician who informed me that he was the reason I had been hired. My first reaction was how? He further indicated and brought news-paper articles where he had written challenging the hospital to hire minority. At first, I thought this doctor was a troublemaker but after a few months, I realized that the same pattern was going on. The first realization was when I saw a fax copy of my position that was adver-tised earlier and the salary did not match what I had been given. You see, I shared the same office with the human resources department so I was able to see and hear what was going on.

Later on, I was approached by a worker in the Human Resources Department who indicated she had some information for me. She showed me a document that indicated I was underpaid and told me she felt I should demand more money. I had only been there a few months, so I did not want to start a fight I could not finish. Also, I felt lucky to make a little more than I was making from my last job, and it was certainly not time to rock the boat. My CEO, in the beginning, was good to me in terms of keeping me informed of what he wanted done. He informed me to immediately attend the local NAACP Chapter meeting and made sure the community knew I was onboard. The other important assignment I had was to celebrate the tenth year anniversary of Our Birth Place, which was the first labor, delivery, post-partum delivery center for women in Georgia. Also, the hospital made most of its revenue from delivering babies, so it was crucial for us to put on a big and exciting program. This was my opportunity to showcase my talents.

I immediately took this project seriously and wanted to start on a good foot and show especially the community that I had the capabilities and could contribute to the bottom line of the hospital. I was especially excited about the fact that the former president and his wife were in the community and made an effort to invite them to the celebration. I knew Mrs. Lillian Carter and President Carter were former board members of the hospital and were regular visitors to the hospital, so it was imperative that I included President Carter in the celebration. I had only two months to prepare for the tenth year dedication, so I did not have a lot to time to get ready. I collaborated with most of the OBGYN doctors and invited all the children that were delivered when the center was initially opened. To my surprise, most of the kids were still around and delighted and their parents were delighted to participate. The event was a success, although I could not get the Carters to attend due to prior engagement.

After this first event, I felt comfortable and eager to tackle other marketing and public relations issues I felt the hospital was facing. I felt appreciated by the doctors and a few community leaders but did not realize that my bosses at the hospital still had reservations about my performance. In my mind, their reservation was justifiable because it was my first big event and I had just been employed for three months. All I wanted was their support and encouragement and time for me to adjust and learn about the issues of the hospital. I immediately realized that most of our patients were Blacks and we had very few Black providers. With my marketing background, I decided to conduct a marketing research (needs assessment) to find out why we did not have minority doctors in the community. I contacted a few community leaders, and they informed me it was impossible to find minority physicians who were willing to move to rural areas. I felt with the right compensation, we could recruit physicians who were willing to relocate to rural areas. I prepared a plan of action and presented to my CEO and asked him if I could work on recruiting minority physicians and also informed him that I had spoken with a few patients who indicated they were not comfortable seeing White physicians because they felt they were still been treated poorly.

While waiting to get a response from my CEO, I took on another project to improve the wait time in the emergency room. The hospital was getting a lot of complaints about the delays in the emergency room but nothing was being done to correct the problem. In my investigation, I found out that most of the patients who frequent the emergency room were Blacks, and nothing was being done to make sure the problem was solved. I brought up this particular issue many times in our meetings and was told that most of the hospitals in the nation have the same problem. This was partially true, but the problem was more isolated to the uninsured and the poor. With the attention given to this issue, and the support of the emergency room physicians, we decided to tackle it as a performance improvement project. This basically meant getting several disciplines of the care team whereby we identify the problem and find an appropriate solution to the problem. I headed the team and appointed several members from department of the hospital I knew I could work with. After a few meetings and months, we came up with a solution and were able to put together an effective way of managing the flow out of the emergency room. This process was presented to the board for approval, and we saw our wait time decrease by 30 percent.

Once again, I was feeling good and important that I was given the opportunity to make changes that were beneficial to the hospital but not given the credit for initiating the process. Our wait time in the emergency room was benchmarked among most of the state hospitals and recognized, but the credit was given to the CEO. This was okay with me, but I realized that my name was mentioned as part of the team, but was not given the opportunity to make a presentation or be part of the presentation. I was responsible for passing out the presentation documents but could not speak. This was not acceptable to me, but I did not have the support. After one of the presentation, I informed one of the doctors that I initiated the ER wait time process and was upset that I was not allowed to be part of the presentation. He indicated that the processor still considered me a new employee with no hospital experience. I wondered whether that was the case or was it because they had never had a Black woman who could speak up for herself and was getting the work done? As much as I wanted to believe that I was being treated equally, I still could not understand why I was running into many stumbling blocks and nothing was being done about it.

My frustration level was building up each day, and the more I watched what was going on at the hospital with patient care and the lower paid employees, I knew I had to get away from the situation. But how? Incidentally, I was more qualified in terms of educational background than all of the senior management team but still regarded as unqualified to work alongside them. My qualifications were questioned, and as a matter of fact, the university I attended was contacted to find out if I, in fact, obtained a master's degree in business administration. Further along in my story, you will understand why this was done. I felt it was done because they were impressed that I had an advanced degree, not knowing that they doubted my achievement.

After nine months of employment as the marketing and public relations director, I was contacted by the director of the Chamber of Commerce to informed me she was impressed with what I had accomplished at the hospital in such a short period of time and wanted to offer me a position with the State of Georgia in Atlanta. I was happy that someone in the community saw and appreciated my work and agreed to go to Atlanta for an interview. I interviewed and was offered the position in Atlanta and decided to turn in my resignation to the CEO of the hospital. Before turning my resignation, I mentioned to one of the employees in the Human Resources Department of my intention, and she informed me that a position for the vice president of external operations has been advertised and she wanted me to inquire about it. She also specifically told me not to tell anyone she told me about this particular position because administration did not want me to know about this position. They felt I was trying to change the focus of the hospital by questioning their practices toward the poor and Blacks and could not understand why we still only had one Black director with over thirty-five departments. To confirm that this position was available, I checked with a friend of mine that worked in the human resources department. They indicated that I look in the fax room for a copy of the position and also, as a matter of fact, showed me the paperwork which included the salary and benefits. At this point in time, I was too afraid to inquire about the position because it would have gotten my friend in trouble and I did not want to put her job in jeopardy.

I turned in my resignation and was ready to move on. I went to see a few of my loyal friends in the community who had helped me to inform them of my departure and explain why I was making a career change. This was not accepted by a few individuals who indicated I had made several changes in the short period of time I was around. They insisted that I stay at the hospital, but I told them, at my position, I was not able to make any changes and had heard and seen enough. I also realized that I did not have the support to make any changes, but with a position in senior management, I might just well have the latitude to make changes.

I contacted my husband to tell him the news that I had been offered a position as vice president of external operations and given two hours to make a decision. I also informed him of the several conditions associated with the offer which were (1) move to the community within six months. He asked me if I wanted to stay at the hospital after all the problems I had been having. I indicated to him that it was a senior management position, and I would be able to have some form of control. He further told me it was my decision to make and furthermore asked me what the compensation was. I told him $65,000 per year. He felt it was not enough, but from my perspective, it was more than what I was making at the time. I knew the position itself was more impressive than the compensation. I made another call for advice from a loyal friend whom I trusted and had experienced working in the corporate environment. He told me to take the job and work on the compensation later. Time was running out because it was about three hours since the job was offered to me, so I had to get back to the hospital and give my boss the response I knew he did not expect, which was accept the job.

The look on his face was one of disgust. He then asked me if the salary was okay, and I ignored his question and rebutted with a request of mine, which was asked to appoint someone to take over my old position of director of marketing, working under my direction. I wanted to appoint a young fellow that I knew deserved the position and would never have the opportunity without me, being a minority. There was a brief silence as to say, "How dare you come in here and make any form of request for the position request." It was known at the hospital that my boss was a tyrant, and no one, even the doctors, addressed him by his first name. It was always Mr. A.

I was holding my breath because I knew I was treading on dangerous ground at this point. The beauty of the matter was that I had nothing to lose. The pressure was on him to keep me at the hospital for political and public relations problems. To my surprise, he informed me that he accepted my request and welcomed me to the position, and accepted my request to promote this young man to a middle management position under my direction. To be honest, I felt that my boss was happy to see me assume the position. He looked at me with a straight face and informed me he created the position for me and commended me for my education and for getting in a senior management position in such a short period of time with the organization. Needless to say, I knew that this particular position had been advertised secretly, but I kept my mouth shut. For a second, I could not understand why the compensation was 35 percent less than they were willing to offer for the position, but the time was still not right to complain. A part of me believed him and felt truly blessed to take up of responsibility of handling non-clinic and clinic areas. I knew it would be a big task, but I was ready for the challenge.

The next day, the CEO called a meeting and announced to the staff that I had been promoted to vice president of external operations. A few of the employees looked shocked, especially the Blacks. I guess they were in shock because in the history of the hospital, I was the first Black to have a middle management position, and in less than a year, I was promoted to vice president, and most importantly, part of the senior management team. I was on cloud nine and could not believe that I will be the one Black in all the meetings and actually be an officer of the organization. It was bittersweet, and I had to give thanks to the Almighty. I also had to give thanks to the one doctor who had brought to forefront that there were no minority in the management team of the hospital.

The next day, an article was written about me in the local newspaper announcing my promotion. I received a lot of congratulatory phone calls from all over the community as well as flowers and well wishes from folk who had never talked to me in the last year. I knew I had made the correct decision of accepting the job although I knew the compensation was not up to par.

Even the Black physician who had worked hard to bring realization to the fact that Blacks were not represented in management position was now black-balled. This physician was responsible for my coming to the hospital and had more than enough evidence to show me how the Blacks had been treated. He took time out of his practice to become the President of the local NAACP chapter in order to give a voice to those who were not treated properly. His primary goal was to give the people a voice and promote equality in the community. He spent the majority of the time advocating for both Whites and Blacks as well as treating patients that were indigent or could not get adequate medical treatment for one reason or the other. This particular doctor went as far as creating a newspaper organization, which enabled him to voice out his concerns, and singlehandedly fought for the rights of anyone that approached him. Because of his devotion to helping the community, he was criticized and labeled to be a troublemaker. At one point in time, his privileges from the hospital was put on hold to hurt him financially, but this process did not stop him from fighting for what he believed in.

As time went by and I saw what this one doctor was doing, I felt that another Dr. Martin Luther King Jr. was in our midst. Despite the torment and ridicule he faced, he never gave up fighting. Most of his patients had respect and admiration for him. He practiced medicine the old-fashioned way, by visiting patients at home and taking care of them regardless of their ability to pay. Through his newspaper, he exposed the fact that the executives at the hospital were being paid enormous salaries despite the poor services provided by the hospital. Also, this article showed that most of the Blacks were paid lower than the Whites in great proportions, including myself at an executive level. It became apparent to me that I was compensated less than managers at the hospital with an advanced degree. Obviously, this was not right, but I needed the job and did not have the resources or support to correct the problem. Also I was reminded that Georgia is an at-will state and without an employment contract, I could not sue or dare complain.

As I watched this particular doctor continue with his fight to bring some form of equality to the community, I began to understand why he was fighting. On one occasion, I asked him why he was spending so much time defending a cause he knew he could not do by himself and without support. His answer was, "Many people before you and I have fought and someone has to continue with the battle regardless of the consequences." I could relate to his frustration and understood where he was coming from because as a member of the executive team, I saw and heard firsthand what was taking place in our meetings. I noticed most of the Black doctors went against this doctor either because they were afraid of losing their jobs or did not care about the mistreatment of Blacks. It became apparent to me that most of the complaints about this man was more racial than valid. There were many instances that other doctors had treated patients inappropriately, but they were not disciplined because they keep their mouths shut. As long as this particular doctor voiced his opposition about the lack of equality to the Black race, he was condemned and efforts were openly made to get him out of town. Although this doctor generated a significant amount of revenue for the hospital, it was not enough because he wanted to see fairness and accountability for what was happening. In one instance, an effort was made to halt his admitting privileges indicating that his medical records were untimely, and therefore, could not admit patients into the hospital. I could not understand this because he was not the only doctor on staff that had this issue. As a matter of fact, there were many doctors on staff that had worse admitting records than he and were not even questioned. As a matter of fact, we had a doctor that had substance abuse problems and was disruptive to his patients but was never disciplined. I know this for a fact because I received a complaint from an employee's husband indicating that this doctor had called his wife thirty-two times in a matter of an hour. I brought this matter to the attention of the necessary party but was told to let it go. On another occasion, this same doctor was brought into the emergency room for threatening to use a gun on his family but nothing was done. I was specifically told to not mention it to anyone and keep it confidential. You could clearly see the partiality and the disregard for equality toward the races, but I could not understand why nothing was being done about it. Nobody wanted to do anything about the problem. It was business as usual. It was obvious that everything was not going on the way it should, but no one was ready or able to change the situation. Business was handled as if we were in the '60s or '70s when segregation was something considered okay or tolerable. The only apparent difference to me there was no visual signs of racism but mental signs of racism. I suppose because we all ate in the same cafeteria, used the same bathroom, rode on the same buses, shopped from the same grocery stores, wore almost the same outfits, had schools that had both races, it meant we were accepted. The only difference that was clearly obvious was the division in the churches we attended. In this particular community, Sunday was the most segregated day. It was assumed that I belonged to a Black congregation, but when I informed one of my dear friends that I was a Presbyterian, she said, "I have never heard of a Black Presbyterian." I could understand where she was coming from because after visiting most of the churches in town to introduce myself to the community, I realized the churches were totally segregated. I also realized the fact that my family was in the only Black in our church and were always considered visitors even after a long period of time worshiping. We never felt welcomed because the older members always treated us as visitors or strangers. My faith was one thing no one could take away from me. My parents were Presbyterians, and I was raised as one. It did not matter that my family and I were the only Blacks in the pew or that people thought it was strange that I was a Presbyterian. Being a Presbyterian is all I have ever known. As a matter of fact, my husband changed his faith to become a Presbyterian either to please me or because he felt called to the faith. I was not going to change my faith because of my color or to be accepted by anyone for that matter.

As time and years went on, it became obvious to me that life in the South had not changed. I came to see that although it seemed as if segregation was in the past, it was evident that it was still present and was blatantly used with no consequences. It was apparent that most people I came in contact with did not care or were not willing to accept. It did not matter if you were educated, had money, or religious; if you were Black, you were regarded as dumb or not supposed to receive the same compensation as the other race. Why was this happening? I thought American was the land of the free and there was equal opportunity and pay for everyone if you met the necessary requirement. Why start another battle? In the '60s, women fought the battle of equality in the working place and although the battle is still going on, it is not as bad as fighting the battle of racism. We want to think that people who say they are Christians are basically good, but in my experience at my employment, I was experiencing a shock that was tough to digest. Mind you, this was the late '90s and early 2000s. The thought of looking back to what happened to the Blacks in the United States was one I did not anticipate, better yet had not experienced, so I could not fully understand the impact. I saw it all over, happening to me and my fellow minority employees. I was at the position to help them, but first of all, I had to gain the trust and respect of my superior to be able to do my job.

Even with resistance, I started seeing the situation getting worse. I was getting a lot of unfair criticism from my boss. One instance was when I was accused of leaking executive information to a director. I was called to my boss's office at six in the evening and told to confess about some farfetched information I had leaked. During my inter-rogation, I was put in an office that was about 55 degrees cold. He made me sit in this cold office for about thirty minutes and insisted I sign a document that had been prepared by one of my peers. I was cold, nervous, mad but would not give in to his threats because I knew that he would eventually use this against me. As a lawyer's wife, I was aware of the many stories of police interrogations and how they use psychological means to intimidate or coerce people into admitting fault. After about forty-five minutes, I asked my boss to provide evidence that I had leaked any confidential information out to anyone, for that matter, and also informed him that it was getting late in the evening and my children were home alone. He handed me a letter which basically said I had leaked information to another employee and wanted me to sign indicating I was guilty. He told me I could take the letter home, read it, and sign it before coming back to work. I felt I had not done anything wrong, so I refused to sign the letter, but would be glad to have an attorney look and the content and advise me on how to handle the situation. My last remarks to him was that I had been informed by my husband never to sign any document without reading it first.

When I was finally let go, my hands were ice cold, I was shaking like a leaf and had to sit in the hallway to gain my composure. It was the worst day in my professional life, and I did not want it to happen to me again or anyone for that matter. My journey back home felt like hell because I cried all the way and had a difficult time seeing because of the tears pouring down my face. I kept this incident a secret from my husband for a while because I knew what would happen if I told him what I had been put through. I wondered if this could have happened to the other vice presidents or if it only happened because I was a target or the fact that I was a woman. Why should a human being be treated the way I was? I am a child of God regardless of my gender or color. I would not do this to anyone, not even to people I did not like. I had to pray for this individual but promised myself I must retaliate when I got the chance to.

It was the weekend, so I had time to calm down and come up with a strategy to follow when I got to work on Monday. The first thing I did was confront the individual I was told make the complaint against. To my surprise, I was told another story which was different from what my boss had accused me of doing. Also after talking to the person who wrote the disciplinary letter, I found out that he was asked to write the letter in the event I confessed. I asked him why he wrote the letter without talking to me first, and he indicated he did what he was told to do. This I know was true because as I indicated earlier this boss was fearless and ruled with an iron hand. Nobody could stand in his way, not even the doctors. Everyone, in my opinion, was afraid of him, even the board members who were supposedly his bosses. How can one person be so ruthless? Who died and made him God? I asked myself that on many instances. I knew his day of judgment was approaching sooner. Eventually, he was going to slip and the fall was going to be a big one. God was watching but taking too long to do something about it, in my opinion. My father always told to weather the storm and fight the battle to win not lose. I was fighting a monster that was too strong and had a lot of soldiers behind him, and I was basically alone, looking from the outside. No one wanted to hear about the problems or treatment employees were facing or do anything about it. For one reason, the hospital was in the black and that was the primary concern of the board. Also it was the third largest employer of the county, therefore keeping the local economy stable and viable was what was more important. Criticizing the hospital was not in anyone's best interest. Everybody and their family wanted to work at the hospital, and it was a known fact that the boss at the hospital was the most powerful person in the community.

As I settled down with my new role, my primary focus was to meet with the employees I was responsible for and come up with an action plan on how, together, we were going to stay profitable, because this was the main focus of the hospital. The employees were not considered important, in my opinion, and the morale of the employees was very low. This was evident because of the many complaints the hospital about their care, and to my estimation, it was because the employees regarded their job as a means to make money and not care for the patients. I wanted to tackle this issue because I knew better employees would result in patient satisfaction. I also felt that the doctors did not support the hospital and were also part of the reason the hospital was not attracting patients that would normally seek care from our hospital. Most of the patients that frequented our hospital were those who did not have a choice either due to financial or transportation.

Being that I wrote my graduation thesis on managing interpersonal conflict, I felt this would be an opportunity for me to present a plan of action to my boss and the board on how we could work with the employees and doctors to improve our customer service rating and therefore increase our marketing share in the ten counties we served. On a Monday night, I was excited to present my plan of action but did not have the opportunity because I was told I could not address the board. The only one who could address the board was the CEO. This was the rule, and no one could challenge this way of thinking. The only other way for me to get to the employees was to take matters into my hands and meet with as many employees as possible. The way of doing this was to walk the halls, meet and greet most of the employees in their work areas. I had to manage by example and gain the trust of those who worked directly under myself and those who reported to the other vice presidents. I was the only vice president who ate in the cafeteria and showed up at every hospital event to handle tasks that were not part of my job description. I felt that there was no task that was not appropriate for me to do even if it meant picking up a broom and sweeping the entrance of the hospital. I wanted the employees to know that we were a team, and as a team, we could accomplish anything. My message was regardless of your title or job, we all had to work for the same goal which was having a successful hospital. The patients, doctors, employees came in that order. The patients paid our bills so they were the ultimate bosses, and without their loyalty to us, we could not have a hospital. For some reason, I was the only one thinking in that manner. I did not appreciate the class differentiation because I also felt God treated us equally regardless of our educational or monetary status. For some reason or the other, my peers felt it was beneath them to mingle with the other employees and treated anyone below their position as second-class citizens. I did not feel comfortable about the way I witnessed the treatment of other human beings because I felt it took all kinds to make a better world. Each of us has their own talent, and everyone has their own specialty which makes us unique in our own way.

During my third year as a vice president, things had not changed that much because I was still seeking for permission to implement any changes to the organization. I could not even manage my areas of responsibilities because I was always asked to consult with my peer who had the same title as myself. The funny thing about this was that each senior management employee was required to have a master's degree, but I was the only one in that position with a master's degree yet was not regarded as qualified to make any decision. I always felt left out because of the way I was handled professionally. I had a degree in marketing and management but had to consult with out-side consultation, and 90 percent of the time, they agreed with what I suggested. I did not understand why we used consultants when I was capable of doing the work or assignment the hospital was paying for these consultants. I had an issue with this practice and wrote an email to my boss indicating that my department was losing a lot of revenue using consultants. The answer I received was that I was not experience in hospital management, therefore, they had to bring outside consultants. It did not make sense to me because the consultants were relying on information from me to make their suggestions to the board. The consultants were there to make money for their companies and had to tell the hospital what they wanted to hear. I knew I could not win the battle with the consultants and the hospital, so I decided to work with the consultants because they needed me to give them the information they needed to make their analysis. This process went on for a while, and I became friends with the many consultants that were hired to work with me or work against me.

The next big project was to dedicate the expansion of the hospital. This was quite a big event, and I knew I had to bring out everything thing I had to showcase the expansion of the hospital. It was a twenty-million expansion, and one of the biggest expansions a hospital of our size had done. As chief marketing officer, I was in charge of the event, but as usual, every suggestion I made was reviewed and scrutinized by folks I felt were not qualified to give me advise. My boss' wife was asked to assist me in the preparations because she was considered an expert in event planning. I am not sure where she obtained her degree in event planning, but it was not a good idea for me to go against my boss's wife. To my surprise, she had a lot of information to offer and was very helpful indeed. She was well-informed about the hospital and knew quite of dignitaries that she suggested we invite. It was quite pleasant to work with her. By the way, she was not a hospital employee and was not paid for the countless hours she gave in the preparation of the event. One of the most important people we needed to invite to the event was the thirty-ninth president of the United States, President Jimmy Carter, who was a frequent visitor to the hospital. Needless to say, we had to send a formal invitation for the president to the Carter Center before including him as a guest speaker on the program. This was going to be an elaborate event, so I was open to the fact of getting a lot of help to put this event together. I knew I did not have the clout to do this on my own, so any help was an advantage to me. It was a twenty-million deal, and I am sure my boss did not have to want a freshman as myself to take control of the event. This was a career move for him, and I, too, understood his reluctance to put me solely in charge of the program.

The preparations began with all the parties involved, including the builders of the hospital and the entire executive team to make sure we covered all the necessary details. A marketing consultant was hired to assist us with the program and direct the event. As the lead person for the hospital, I felt it was my duty to take a lead position in the preparations, so I made sure during the meetings my input was heard. We collectively came up with a program that I felt was reasonable for the program and submitted it to the board for approval. The first draft was rejected because my boss did not want any female on the program. I felt that I should be the mistress of ceremonies because I was the marketing executive and that was the appropriate thing to do. Needless to say, it was not a good idea. I wanted to find out why but was advised not to argue with the boss, so I left it alone. The dedication of the new addition of the hospital was more important than me getting heard so I decided to take a back seat and work with what I had at hand.

The day of the celebration was getting close, and the final preparations had been approved by the boss and the board, but at the last minute, I was told that I had to make a few changes to the program which was have the speakers not face the crowd. I felt this was wrong because the crowd were our guest and they should not look at the back of the speakers. The crowd were the customers, and I was not going to have them treated as not important regardless of who we had speaking. From an etiquette point of view, you never have a speaker not facing the crowd when speaking. I was further told that the people were not that important especially with the attendance of people like the former president in attendance. About thirty minutes before the program started, I realized that my boss' wife had the chairs moved and changed the seating that we had spent all day arranging. When I came for a final check-up and realized that she had all the sittings changed, I immediately called on my staff and some of the folks that had gathered around to help change the sitting. This was not a good idea to fight with the boss's wife, but I knew I was doing the right thing and was ready to face the consequences. I had spent the last month working twelve to fourteen hours a day preparing for this event. I worked harder that anyone in the hospital, and without the support of my peers, I was determined to show them how an event should be handled. To be fair, I had the support of the consultants who thought I knew what I was doing and supported my inputs and actually agreed with my suggestions and direction. In the back of my mind, I could not understand why the people I work for did not have the confidence in my work but knew that it was the norm. After all, I had experienced this type of behavior toward me since I started my employment even with the previous employer.

As I got the call from the Secret Service that the president was on his way, I became very nervous and knew it was time for the show to begin. I had been told to let my boss know when the president was on his way so his wife can meet with them and walk them to a holding place before the program started, but I knew this was my job and I was the one who needed to take them to the holding area, so I did not tell my boss of the president and his wife's arrival. I had done most of the work and as the marketing executive and the person in charge of the event, it was my right to welcome the president and his wife. It was time for me to make a stance and not be treated as a second-class citizen. I was not allowed to speak at the event, so this was my only opportunity to take some credit for my work, which was walking with the president. This felt so good, because I thought who in the world would have thought a little girl from Africa will be walking next the thirty-ninth president of the United States of America? What would my parents think of this moment? It was one of the greatest moment of my life, and I was not going to let anyone take it away from me. This was an opportunity I deserved because I had worked hard to put this event together, and regardless of my color or gender, I was going to be seen and heard.

As mentioned earlier, the program speakers were all men as requested by my boss, but I took it upon me to include myself on the program although my name was not listed on the program. I got onstage at an appropriate while waiting on our local congressman to introduce the president and said a few words acknowledging the president. As I made my way on stage, I could see the tension on my boss' face but did not care of the outcome. The look on my boss' face was one of shock and disgust, but you see I had an answer planned in the event I was asked why I felt necessary to speak when I had been told not to. I felt vindicated and knew a lot of people were proud of me. I felt I had left my people happy because I knew they had never seen a woman, much less a Black woman, stand in front of them and actually speak in font of White folks. During the reception, I had the pleasure of speaking with Mrs. Carter and had a chance to find out more about her institute for caregiving which I eventually had the opportunity of working with (See exhibit). I had gained the respect of a lot of people by my actions. It felt good, but I knew this was the beginning of the end. After the program, to my surprise, my boss commended me for an excellent job well done but still attributed the success of the event to the consultants.

As the months and years went by, I felt committed to my job and was getting a lot of recognition from the community. I was being asked to join a lot of organizations in the community and asked to sit on several boards. This was a good thing, but at the hospital, I still felt my contribution was not accepted. I still felt as an outsider because most of my suggestions were not listened to or taken lightly. I continued to work on my programs securing enough funding for a particular program (The Rural Health Program for Migrants) which had very little funding and was going to be canceled. This program was one that took care of migrant workers in more than six counties and was vital for us to continue because it helped the hospital prevent the emergency room cost. From my research, I proved that most of our emergency room cost were from the migrant workers coming in for care, and because they did not have the ability to pay or insurance, we had to write off most of the costs. By providing services to this particular segment, we were able to reduce our cost. My goal was to ask for more funding for the program, which I did, and lobbied for the program not to be eliminated. I had a wonderful nurse practitioner who was one of the most intelligent women I had met in my professional life. She had worked with a lot of underprivileged people and had devoted most of her life helping the needy. She was passionate about her patients and taught me a lot about working in the rural area. She was an inspiration to a lot of people and actually was the reason we had a successful program. I was fortunate to have her as my director, and although the credit for having a successful was given to me, it was because of her commitment and dedication that the program worked. We started with a budget of $40, and within five years our budget was increased to over $400,000. Despite this success, the program was slated to be canceled because my boss felt we were losing revenue and the community (leaders) did not appreciate the hospital treating migrant workers. I could not understand the logic because we were saving quite a bit of money not using the emergency room which was where most of the uninsured were coming in for treatment. I felt it was a political problem because there were other programs that were not generating revenue but were not slated to be cut. I knew I had a battle on my hands to fight for the program not to be cut, so I solicited the support of the local farmers, who were the employers of these people, and challenged them to fight for the program. This was not an easy battle for me because the purpose of the hospital was to generate income and I did not have the support of the board.

My next step was to call on our local congressman who needed support from the farmers, and I knew that, politically, he would agree with keeping the program. In my humble opinion, I felt the only reason we wanted to eliminate this particular program was because the majority of the patients were poor migrant workers and a lot of the so-called patients did not want the patients in the hospital. They felt these patients were poor and bringing unknown diseases to the hospital although there was no research that had been conducted to show this was the case.

Because I could prove that the hospital was not losing money on this particular program, I took it upon myself to invite the State of Georgia to further explain why the program was beneficial to the hospital and why it should not be eliminated. Besides, our funding was the highest in the state, and our program was a model program for the state. Also, this program was very beneficial to the local economy, and we did not only treat migrant workers but low-income individuals. With the support of the congressman and the local farm-ers and my nurse practitioner, the program was funded for another year. This was another win for me, but I was getting frustrated about fighting for programs that I knew were beneficial to the community as well as the hospital. They were a lot of services and programs the hospital had that were not generating income, but they were not on the chopping board which I could not understand because I was being questioned all the time about the programs and services that I was in charge of overseeing. Each time I brought up the fact that other services or programs needed review, I was told that I did not have enough hospital experience although I was the only person on the management team with an advanced degree and just as much experience as the others.

The days, months, and years became more trying because I felt left out of decision-making on areas that I supervised. Moreover, I had to consult with my peers on what to do because I was not allowed to make any decision even in areas I felt I had ample experi-ence in. This situation bothered me a lot, but I had no one to turn to. It was obvious that everyone was looking the other way and would not understand why I was complaining or why I felt I was being treated unfairly. I say this because, at the end of one year in particu-lar, the hospital had generated enough revenue to give the managers and executives a bonus. This bonus was based on a percentage of our salaries. When I received my bonus, it was considerably less than what some of the middle managers had gotten. The reason I knew this was because I was told by an insider that my bonus was less than the others, and when I questioned the proper authority, I was told I should be happy with what I received. There was always an explana-tion of how I have been treated, and I knew I was fighting a losing battle. No one was willing to speak up against the management team, not even the board members that were over the hospital. As a mat-ter of fact, we had individuals on the board that I felt I could take my concerns to and at least get a fair hearing or make my situation bearable. I realized that these people were on the board for political reasons and did not care about the effective running of the hospital or made sure that the administrator was ethical. It was a fact that most of the board members were selected by the CEO, and they did as they were told. They did not make any decision that was beneficial to the hospital or the employees. Most of them came to the board meetings to get a good steak and agreed to whatever was presented to them for their approval. As a matter of fact, I was not allowed, under any circumstance, to speak to a board about hospital affairs. It was to all of the management team not to communicate with the board members about hospital business. Board members were regarded as God and nothing was to be said to about anything, and if questioned by a board member, you were to let the CEO know immediately. A separate table was always set for some chosen individuals during board meetings because we were not allowed to sit by them during meetings in the event we start a conversation. On one particular night, I had to give a presentation, so I felt it would be okay to sit at the main table with the other presenters. This was not received well because the next working day, I was called into my boss' office and interrogated on why I sat by a board member. I was not the only one who was given this specific instruction not to sit by the board members. Particular board members were regarded as kings, especially the chair of the board. It was common knowledge that this king made all the decisions regarding the management team, and my boss treated the chairman as God.

One particular night, during our board meeting, the king spilled sauce on his tie, and one of the servers had to run out and get soda to clean his tie with. You see, just Blacks served dinner during the board meetings, and they were always so timid and acted as if they were back in the slavery days again. Another very important observation was that although our organization was a non-for-profit and the meetings had to be published in the newspaper, it was seldom done because we did not want the public to know what was going on in the meetings. In the event we had to give notice of a meeting, it was done reluctantly, and everyone was excused before the real meeting started. This was called "going into an executive session" and most of the executives were excused as well. It was a common practice to have me attend the meeting each time we had a visit from the local NAACP organization to give the impression I was of some type of importance although I was instructed not to relay any information to them. At these meetings where the NAACP was present, I was also given the opportunity to make a presentation or get called upon to answer questions to give the impression that I was involved in the decision-making of the hospital, which was to the contrary. As a loyal and devoted employee, I always defended the hospital and did as I was told. This bothered me a lot, but I needed my job, and in the back of my mind, I felt it could only get better, since the hospital was using me as the go-between with the NAACP. The main focus of the NAACP coming to the meeting was to make the board aware of the salary discrepancy they had with the minority employees. The organization had gone as far as requesting the salaries of all the management team and printed them in the local newspaper. It was obvious that I was being paid less than White employees that were in middle management, but the board still did not find it necessary to make any changes to my salary. Even after a compensation company hired by the hospital to evaluate executive salaries indicated that my salary was below the average norm, the board and the CEO found many excuses to not compensate me based on the recommendations of the compensation committee. The only recourse I had was to let the president of the NAACP know that I was being paid less and gave them the necessary documents to indicate the findings. I knew that eventually with the involvement of the NAACP and the community, it would be changed.

As time went by, the doctors had issue with the hospital CEO because of his control and started demanding for his resignation. Unfortunately for the doctors, the CEO had the support of the chairman of the board who, in my opinion, had control of the hospital and basically made all the decision. As a matter of fact, one of his family members built a physician's office in a community that was saturated with doctors and did not need another doctor. Although extensive research was done to indicate that this was not a good idea to use hospital revenue to add to a community that had too many doctors, the board approved the construction of a new clinic and went ahead with the construction. Yet my programs that needed revenue allocating and were beneficial to the hospital were slated to be cut. How does this make sense? You can judge for yourself and see that we were giving a contract to a board member to use hospital revenue for a project that was not needed yet refusing to invest in a program that was not only generating indirect revenue for the hospital but taking care of a needed population. All I can say is that the project my programs were slated for were for the poor and that was not a priority for the hospital. It was obvious that no matter what I did or tried to do, I was not getting any support from the organization. All of my frustration was expressed to anyone who would listen, including community leaders and the board members I felt would listen. I knew I was treading on thin ice, but I also knew the only way I could be fired was to do something illegal. This is not to say they did not try to frustrate me to resign or just walk out and leave behind a lot of people who depended on me since I was in senior management position.

As the Lord would do it, I continued to struggle through the daily torture of dealing with my boss and working under extreme stress without the support of anyone except those who worked under my supervision. I knew I was doing the right thing so I continued to pray and hope for the day that someone would listen to me or accept the fact that I was a woman who had an exceptional background in health care and capable of making important decisions for the viability and survival of the hospital.

It was very important for me to keep my religious beliefs intact to continue to work in the environment I was in. I had to stay focused because without God, there was no way I could continue to deal with the pressure and the people who were in charge of running the hospital. I had to tell myself every day that things had to get better. Each day, I had to pray and remind myself that things had to get better. As a child, my father taught his children that if you believe in God, you will make it regardless of how long it took, but on the other hand, I was beginning to doubt if there was really a God because it was taking too long for changes to happen. I could not understand why and how one person was making so many bad decisions and getting away with it. You have to understand that I was still thinking in the past or how I was brought up to believe that all people are treated the same and also for the fact that I had been told by my parents that I was in the place or surroundings (the White man country) where people were treated the same. I could not come to grips with the fact that I was a Black woman, and no matter what I did, I was a minority and would never be treated equally. No matter how educated, how well I dressed, how well I spoke, or what type of car I drove, I was *still colored.*

A few years later, the hospital was able to add on a twenty-million addition to the existing building. This new addition brought in more services and doctors to the hospital and elevated the authority of the existing administrator and board members. This obviously gave more power to the ruling class and did not change any of the discrimination that was going on. The goal of the hospital, in my opinion, was to compete with the larger hospital around us and eliminate a lot of the services rendered to the poor. To be able to be financially stable, we brought in more consultants and had to eliminate services geared to the poor. Most of the programs that were offered to the community, such as assistance to the local school systems, pre-natal care to the poor, or rural health care to the migrant workers were seriously cut or eliminated. At this junction, the hospital implemented a program whereby folks were being sued for not paying their bills. This became a public relations issue because a lot of the poor (Blacks) were having liens put on their homes or wages garnished because they owed the hospital money for medical services rendered. I felt this was not a good idea because the less fortunate people who could not go out of the services area to seek medical care were being punished, and those who could afford insurance and had the means to pay for their services were not even using our hospital for services. This was obvious because in a survey conducted by the marketing department to find out about our market share, it was noted that most of the patients with insurance did not seek medical care at our hospital. My argument to this point was that our services were geared to the poor but we did not take care of the poor because about 70–80 percent of the patients that frequented our hospitals were Blacks. As a matter of fact, most of the workers in middle management or senior management were Whites. As you can see, the decision makers were Whites and did not care about the patients because obviously, the majority patients were Blacks, and as indicated, they could not afford to travel out of the services area and had no other choice but come to our hospital for their basic medical care. In the midst of all of these issues, the hospital was getting ready for an elaborate opening of the new addition to the main campus. This was a twenty-million-dollar addition, so it was going to be a big celebration for the community. Also this was a project dear to the heart of the CEO, and he wanted the entire South Georgia to see his masterpiece. The guest list was to include the thirty-ninth president of the United States and every one in south Georgia who had a public service position (senators, councilmen, mayors, community leaders). Though this affair was to be led by the Marketing Department, led by me, I was instructed that my role will be limited and had to work with a consultant out of Atlanta. I had no choice but to go with the program and work with the folks out of Atlanta. I knew I could not fight the big boys, so I agreed to work with the consultants although I knew I could handle the task with the staff I had. The was no reason to spend extra revenue to hire a consultant when I had enough staff to get the job accomplished, but once again, I was not good enough to handle such a big event despite the fact that I was the executive in charge of this particular operations.

The plans for the dedication was complete and all the guest speakers were confirmed, with the thirty-ninth president of the United States as the main speaker. One other important fact was that we could not have a lady on the program, as instructed by my boss. Since I was the organizer with the help of the consultant, I felt it will be only right for me to be the lead person to oversee the program. Also, as the executive in charge of marketing, it was my duty and job to make sure the program was done successfully. After all, I was hired as the marketing executive, and technically, it was my assignment to conduct and direct the dedication of the hospital.

As we were finalizing the program for the dedication, I was informed by my superior that I had to work with an individual who was not even an employee of the hospital. The fact of the matter is that this individual was the wife of the CEO and was considered an expert. I did not have a problem with this particular lady because she had been always kind and respectful to me in the years I had been at the hospital. I considered her a friend and was willing to work with her but did not know that she would want to have me work for her instead of with her. Our disagreement happened when she wanted the speakers to give their backs to the guests. Her reasoning was that the speakers were dignitaries and were not supposed to face the crowd. I felt the guests were the ones who keep the hospital financially stable and had to be treated as customers, not treated as if they were not important. As a matter of principle, the hospital had always treated our patients as secondhand citizens, and this was my opportunity to make them feel they were just as important as the president of the United States of America. It was the practice of the hospital to treat our patients and employees as secondhand citizens, and this was my opportunity to change that feeling. I was always told that you could always ask for forgiveness afterward, and I was going to take that and run with it. I knew I could not be fired on the spot for something that stupid, and knowing me, I would make a bit deal of it.

The next problem, as we were preparing for the arrival of the president and his wife, came to light when I was told that I could not walk them in to the hospital but the wife of the CEO had to do the honors. Well, it did not sit well with me so I instructed my assistant to notify me as soon as the Secret Service personnel were at the hospital grounds so I could talk to them and let them know that I was the person in charge and should be the one to escort the president and his wife to the waiting area. To make things more uncomfortable, I was approached by one of my peers to let me know that the boss was getting more irritated by my actions and wanted me to follow his wife's instructions and do as I was told. At this point in the game, I did not have time to waste or worry about the consequences. My main objective was to get the show on the road, and as a perfectionist, I knew what I had to do. I knew I was in trouble already, so I ignored him and made sure the necessary preparations were intact.

To be safe, I made contact with the Secret Service personnel and informed them that I was the person responsible for the event and would appreciate them letting me know when the president arrived.

About thirty minutes later, I was informed that the president was on his way, and I immediately made my way to the front entrance of the hospital to await their arrival. As the motorcade arrived, I walked with the Secret Service lead person to welcome the president. I lifted my head up high, and with much confidence, walked as close as I was allowed to welcome the president. To my surprise, the president, with his usual trademark smile, extended his hand and greeted me with grace. I felt unusually important at this point but knew I had to face my boss punishment later.

As I escorted the president and his wife to the waiting area, I could see my boss wife at the corner of my eye, very furious. I did not care at this point in time because I knew a lot of people were depending on me. Being calm was my main objective, and no one was going to upset me or divert my attention on such a big day for the poor individuals who depended on the hospital for their medical services. The celebration was really for the rich people and not those who needed the services. We spent a lot of money for the celebration but could not spend would not give the poor a break on their medical care expenses. Needless to say, the celebration had to go on.

As the program began, I had been instructed not to speak or get on the podium. My boss—who was not a great speaker, by any means—was to be the first to speak, which was the proper thing to do, but I felt I had to make an introduction to begin the program, which I did. When I walked up to the podium, I immediately recognized the president and his wife and then handed the microphone to the CEO for his comments. I was very pleased with the turnout and felt vindicated because the program was well-organized. I made sure employees of the hospital were involved in the celebration. I had employees from each department represented in the celebration. The way I used these employees was to have them participate in giving tours to the guests and visitors. This gave a lot of the employees who were always in the background a sense of belonging and pride for an accomplishment that was done by each and every one that worked for the hospital. This addition was not only done by senior management. It was a collaboration by a lot of people, and I wanted the community to see that it was a team effort, and also for the fact that a lot of our employees needed to be part of the organization's success. The morale of the employees were at an all-time low, so getting them involved with the dedication of the new addition was a priority for me. I understood how it felt being left out, so I was determined to get as many employees involved in the celebration. I made sure I had a representation of both White and Black employees involved in the celebration. As I mentioned earlier in the book, Black employees worked in the areas such as kitchen and environmental services, so they were never seen in the public or were never given the opportunity to be part of the celebration of the hospital or even made part of the organization. They were there to clean and cook, and that was their place in the organization. I think most of them were comfortable in that role and did not know any better. It was also told to me that the Blacks did not have any education and the hospital could not recruit Blacks because most Blacks did not want to move to rural areas.

After the dedication, the elite guests were taken to the administration club house for a lavish reception. At the reception, I was congratulated by Mrs. Carter for a job well done. As usual, I could not have done the job without a lot of assistance from the entire employee base. At the end of the day, I was happy with the outcome of the celebration, but knew I had to deal with the ramification of not letting my boss' wife run the show the way she wanted to. At this point, I was ready to defend myself because the celebration went well and the turnout for the celebration was tremendous. I knew the Lord was watching over me, and I was prepared to fight at any cost if my boss confronted me with any negative thoughts.

When I arrived at work on Monday, I was elated to read the press coverage of the celebration and noticed that I had a picture of Mrs. Carter and myself in the front page. I knew I was in more trouble because, as a rule, I was not supposed to appear in any pictures, much less be on the front page of the local newspaper.

As I walked to my office, I was approached by one of my peers and was asked why I had my picture in the front page of the paper. He informed me that the boss was upset, and I better have an explanation why my picture was in the newspaper. I tried to explain that

I had no idea why it happened but knew God was watching over me and wanted my picture posted in the paper for the people of the community and others to see what a good job I was doing. Regardless of the press coverage of the celebration, I was still afraid of what my boss was going to do to me when I met with him to discuss the event. As I approached his office, the first thing I wanted to know from the secretary was if I was in trouble. As I walked into my boss' office, I felt a weakness in my legs and said a prayer and asked God to protect me from him. To my surprise, he did not mention anything negative or positive about the celebration. I was so shocked that I starting thinking my mind was playing games with me. We talked for about five minutes, and I was out of his office.

As the months went by, I started thinking of ways that I could help the hospital improve its revenue given the fact that I had a master's degree in business administration. Most of the areas under my supervision were non-revenue generating but I knew I could start programs that would generate income for the hospital. One of the problems that the healthcare industry was having was a shortage of nurses. This was a problem that had been discussed in our management meetings, and I wanted to be the one to come up with a solution. I took it upon myself to discuss this issue independently with the chief nursing officer to get her take on my plan of action. It was a known practice at the hospital that you do not tread on someone's department, especially the male vice presidents. Because the chief nursing officer and I were the only two women on the executive team, we seemed to work together. I was very fortunate to have another female vice president as a peer who had a lot of experience and had been very supportive. She gave me the support I needed to effectively manage the clinical areas under my supervision, so it was important to me with her to get her support and advice. At this point in time, she was also getting a lot of criticism about reducing her staff and cutting staff, which in my opinion, was not feasible. Although she managed two-thirds of the budget and generated almost half of the revenue for the hospital, she was not given the support she needed from the CEO but had the support of the doctors and the community leaders. Although I felt as if I had a supporter, being a woman as myself, I knew that she possessed something I did not have, being a White lady and a woman who grew up in the thirties or forties and was well-aware of how Blacks and women were treated in the South. In all honesty, she was very kind to me and gave me as much help as I needed in terms of running the clinical areas which was something I was not totally comfortable doing but was qualified and capable and had personnel who were more than capable of running the program.

After further research on the shortages of nurses in the entire healthcare industry, and after carefully studying the problem with experience consultants in healthcare, I took it upon myself to write a letter to the Governor of Georgia, informing him of the problems hospitals were having retaining nurses and suggested recruiting nurses from English-speaking foreign countries. The reason for recruiting from the English-speaking countries was due to the fact that some hospitals had been recruiting nurses from Asia, and most of the nurses had problems communicating with their patients. I was confident that I could build up a pool of nurses from my home in Africa because as a little girl, my father provided housing for 90 percent of the students who attended nursing school in my home town. Also, my older sister was a nurse and she was willing to recruit and train nurses for me to import to the United States. This was a brilliant idea because at this point in time, the hospital was using agency nurses, and it was costing the hospital not only a lot of money but the nurses could not communicate with the patients because of the language barrier. There were other factors, such as problems with the nurses understanding the doctors' orders and the doctors expressing their frustration dealing with these non-English speaking nurses.

After putting my plan together, which included a cost-and-savings analysis for recruiting nurses, I presented it to the executive members of the hospital which included the CEO and the rest of the senior management team. To my surprise, it was not only ignored, but I was told I did not have the experience or foresight of what was going to happen in the industry. I could not understand the comments because the shortage of nurses was not only a problem in Georgia but through the country. As time went on, this particular problem became worse, and we started using more agency nurses and saw the hospital's revenue drop.

Since I did not get the support from my superior to tackle the problem of the shortage of nurses, I decided to come up with another bright idea of recruiting minority doctors to our community. This was thought of because as I visited and talked with a lot of the patients that came to our hospital, they indicated that they would like to see the hospital recruit minority physicians. A lot of the younger patients wanted to see doctors that they felt understood their problems and felt the physicians we had on staff did not meet their requirement. The older patients did not particularly care about who they saw but were in agreement of having a choice of whom to see for their medical care. The other concern that was related to me was that the White physicians did not take the time to listen to their concern and it still felt like the days of slavery when patients could not ask the doctors questions about their care. Basically, it was related to me that they felt they were getting substandard care. This was apparent because most of our patients that came through the emergency room, had to wait up to ten hours to see a doctor, especially if they were Blacks, and from the numerous complaints that came to my attention, I had to make an investigation. When I did my investigation about the complaints and found out that a few physicians were not treating certain patients the way they should, I was immediately told not to say anything further about our emergency room wait time or how the patients were being treated. The problem was that the emergency room physician at that particular time was a board member and had been with the hospital for many years and could do no wrong. He owned the emergency services and was leasing his services to the hospital so he did not care if the hospital's revenue was falling. He had no interest in the hospital viability, and as far as I know, he was not to be questioned about his management style. I knew I had to come up with another strategy to tackle the wait time in the emergency room so I solicited the advice of a friend who had some clout in the community and was a nurse practitioner. I asked her to look at the emergency room issue as a performance-improvement issue and form a team from departments to tackle the problem as an improvement project. I also wanted to be put on the team but at a very low capacity because I knew if my boss knew I was heading this project, he would not allow the project to go forward. To my surprise, the project was launched and the same issues I discovered were still happening, and when the findings were brought to light and the recommendations implemented, it became an award-winning project for the hospital that was used for survey purposes. Needless to say, this was a project that I had to do using others because I was still not considered capable of coming up with good ideas.

As I reflect on all of these situations, I can only see why now. As much as I saw blocks put in front of me, I did could not see the fact that I was a Black woman and the people around me felt that I was not capable of coming up with solutions. Better yet, now I can believe that they felt I did not have the tools of a human being just because I was Black.

As the days and months went by, I continued to fight the battles I could and dealt with the ones I could not. I made it clear to those I could that I was not going to leave the organization because I had other people who depended on me because of my position. I also knew that with my personality of not bowing down to anyone, I could take the pressure. My bark most of the time was bigger than my words so I used it to scare a lot of people. I heard from the other members of senior management that I was being called names but it did not bother me. I knew that they felt my husband was a lawyer, and I would seek legal counsel if need be. I also, on several occasions, used the fact that I was from the north and that I was not afraid to sue if I had to. For some reason, they also felt that I had a lot of money and really did not need the job to survive. Although things were getting more difficult for me to go to work each day, I had the greatest supporting staff that made me want to come to work each day. These people were brilliant and were very committed to me as well as to their work assignment. One of the reasons for this because I also believed that if you treat an employee with dignity and make them feel as if they are part of the decision-making, they will perform at their highest level. I also made sure that my employees had the fullest access to me as well as feel as if every decision that was made was done as a team, not by myself because I was the vice president. I made sure that I knew each of my employees as an individual and tried to bring out their strengths. I also made sure that the other employees of the hospital could come to me if they had problems because we all worked toward the same goal, which was giving our patients the best care they deserved. This was not the attitude of the other senior managers. They were only concerned about their individual departments and did not care if we all worked together as a team. The business at the hospital was very territorial, and I felt totally uncomfortable working there. It was seldom you see a senior manager walk the halls and greet a patient or talk with employees of the hospital. The atmosphere at the hospital was one of fear and confusion. The morale of the employees and doctors was very low because administration was only concerned with making money, and it was a fact that the senior management team were the only employees with high salary. This was something I took issue with because my salary was lower than some of the mid-management staff. When I complained about my salary to the appropriate authorities, it was ignored. I remember the first year that I was in senior management, the hospital was fortunate to have a positive bottom line which meant middle manager and senior managers were entitled to a bonus. I was excited because I knew I was going to get a large amount of money or something comparable to the other vice presidents. To my surprise, I only received a very small portion and the reason being I had not been in that position and did not have health care experience. When I took up this issue with a board member, I was told to shut my mouth and be happy that I got any money in the first place.

Things did not change for me as time went on, but I was enjoying my work and getting to meet a lot of people in the community who thought I was doing a fantastic job. Also, it was comforting to me to see more services being offered at the hospital. Our physician base was improving, and although I was not given credit for many recommendations I suggested, it was eventually done through consultants, with whom I collaborated with. I realized that the only way to be heard from is to do it through another means, which was voice my opinion to anyone who would listen. It was a known fact that I was not afraid to fight for what I believed in, and I knew that the folks at the top did not believe I would dare tell anyone in the community that I was being treated poorly. It was a known fact that most of the community leaders knew I was a token and also felt that I should be happy to be in the position I was. I remember at a chamber meeting being told by the mayor of the town that I was a token, and furthermore, should count my blessings because he could remember the time when this town did not allow Blacks to walk through the front door of the hospital. He also mentioned to me that I although my salary was lower than individuals who held lower positions than I did, it was more than any Black person had made in this part of the country. I did not know how to take the comment, but I knew eventually this injustice will be brought to light. That same year, the NCCAP started looking at top salaries at the hospital and wanted to know why the charges at the hospital were so high yet the executive team had enormous salaries. They decided to demand the salary of all the executives at the hospital through the Free Press Act, and since the hospital was a not-for-profit organization, they had to produce all of their management salary. Well, the hospital knew there would be a problem if the salaries were disclosed and my salary was below mid-management (or the good old boys') salaries. As this was going on, I was called by my boss out of the blue and told that my salary was going to be increased. I was shocked because I was not due for a raise or had not been told that I had done anything spectacular to deserve an increase in salary. After the salaries were printed in the newspaper, I realized that my salary increase was just a few dollars more than directors that were less qualified than I was and quite less than my peers. I pretended as if I did not know what had taken place. Although I did not have the means of changing the devilish things that were going on, I had my spy on the board that kept me informed of what was going on. I had a doctor on staff who generated a lot of revenue for the hospital and was a board member, who did not like the situation at the hospital but could not do anything to change the situation, who told me to buy my time and wait because eventually, everything will come to light. This board member who felt I was being mistreated, told me he could not help me but would give me the information I needed to sue the hospital for discrimination. Although I thought about taking his advice, I knew Georgia was an at-will employer, and it was not worth me hurting my employment chances with other employers. I was smart enough to know that most employers will not hire you if they think you are going to bring legal actions against them, and your compensation is not going to amount to a lot in the final analysis. For this reason, I had to continue to go through the torment and hope that, as time went on, things will change, or at least someone will see what was going on and make a change.

As the years went by, things at the hospital were not going good for administration. The doctors were getting frustrated because they were not getting their way in terms of salary as well as facing tougher reimbursement challenges. If you have ever worked in a hospital setting, you will notice that doctors are not the easiest people to work with or for. They are very demanding, and everything has to go their way. They are basically arrogant, and for the most part, demand respect even when they do not deserve respect. I must say they are a lot of doctors that do have compassion for their patients but for the most part, most of them give you the impression they know it all. On the other hand, if you treat these doctors as if they know it all and give them the highest form of respect, they will listen to you and respect you. I always found a way to deal with the doctors I worked with because I took the time to know them individually and spent a lot of time with them listening to their concerns or making sure that they received the help they needed in dealing with their patients or their practices.

For some reason or the other, most of the senior administrative staff at my hospital did not think the doctors were an equal partner when it came to running the hospital. For these reasons, my boss started having problems with the doctors, and this was the beginning to the end of his rule at the hospital after a period of over twenty years. When things started getting so bad with the doctors, my boss who had never thought of me as a vital part of the administration, started visiting my office to consult with me about dealing with the doctors. I was not surprised because he knew among the entire senior management staff, I was the only one who had a good relationship with our doctors, and knew for the hospital to be profitable, we had to include the doctors in the decision-making or at least get their input or suggestions on some of the issues that were facing the hospital. As his subordinate, I was willing to help him in trying to bridge the conflict he was having with the doctors. As a matter of fact, I became a spy for my boss. I found myself playing a very dangerous game, which was finding out what the doctors were saying about my boss and going back to him every morning to keep him informed of what was going on. During these meetings with several of the doctors, I found out that they were very upset with the way my boss was treating them and felt it was time for the doctors to replace him. My boss did not think the doctors could get rid of him because he had the support of the board members. Needless to say, without the doctors admitting their patients to the hospital, I knew, financially, it will be only a matter of time before the hospital went under.

The last straw came when my boss fired the chief nursing officer because she had a private party in her house to celebrate the hospital passing a three-day accreditation from JACHO. Fortunately for me, I was not at the party because for some divine reason, I had to get home forty-five miles away from the party to be with my children, and I had been working sixteen-hour day in preparation for the survey. The next morning when I got to work, I was told that my friend, whom I helped get the job as chief nursing officer, had been escorted out of the hospital and terminated. I remember this because it was my birthday, and I felt sick to my stomach because I knew I would have been fired as well if I had attended the party. The Good Lord was watching over me. As the news spread throughout the hospital of the termination of the CNO, the tension was so high, and everyone was talking about what had happened at the party and how the employees were dissatisfied with the boss and wanted him fired. I was called to my boss's office, and at this point in time, he wanted my support about his decision to fire my friend. I did not know what to say but had to agree with him because as I mentioned earlier in the book, he's was the king of the land and no one dares cross him or disagree with him. It was a good day for me because it was my birthday, and as a child growing up, my mother always told me nothing bad can happen on the day the Lord brought you into this world. I spent my birthday thinking of my friend and fellow female vice president but knew I had to protect myself. I knew we all had to fight our battles and looking at the past, I had always fought my own battles. Everyone at the hospital was always too afraid to listen to my concerns, and I was not going to jeopardize my job at this juncture to protect anyone who did not work under my direction or who came to the hospital for medical care.

As the pressure built up on my boss from the Black community and the doctors, I found myself defending the same person that had treated me poorly. I remember years earlier, when I had to fight for my job because I was accused by my boss of submitting false receipts for my mileage. Each time I traveled out of the hospital for a conference or hospital-related events and submitted my travel expenses, it took an act of congress for me to get reimbursed. On one occasion, I was interrogated for almost an hour about my expense report. I was told if I did not agree that I had falsified my expense report, I will be fired. My boss called the human resource director and had him in a meeting with me to confront me about a trip I took with another vice president and indicated that my mileage report was about 100 miles more than the other person, although I lived more than 100 miles away. I could not understand this form of torture, but I stood my ground and refused to say I did anything wrong. It was always a problem for me to get reimbursed whenever I submitted my expenses. Usually, I would call the secretary several times over a period of weeks to find out if my expenses had been signed off and beg her to make sure it was done. The other male senior management staff did not go through this process. As a matter of fact, I seldom got approved to attend meetings out of the hospital except when I was working. To make things worse, when we attended meetings out of the hospital, as a vice president, I was always working while the rest of senior management were partying or enjoying themselves. Although I knew I was being treated differently than the other peers, I eventually accepted my role as a token and hoped for the day when the tables will turn, and pray that my children never have to go through what

I was facing or experiencing. The sickening part of this whole ordeal was that I knew I was being discriminated against, but could not understand why it was happening when we lived in a society where equality was supposed to be. The question I asked myself many times was why and why could I not see the fact that I was a Black woman, and no matter how educated I was I was *still colored*. I always felt that the situation will change and this could not be happening. We all have our vices when it comes to dealing with people regardless of their color or gender. In my situation, I saw my responsibilities being taken away from me. I was like an object to be viewed and not heard. I was confident in my abilities and knew I could make a difference in the organization could not get pass those who were in charged. Years back, I was so passionate about finding ways to eliminate obesity that I started a walking club at the hospital. I, also with the help of my nurse practitioner, wrote a grant to address obesity in our local school system. Johnson and Johnson as well as Johns Hopkins University looked at our findings and gave us a grant to work with fourth and fifth graders in a local school. Today, Mrs. Obama is vigorously working on childhood obesity, and this is the same issue I had been working on for years and no one would listen to my ideas. To keep my sanity, I decided to start a choir with the employees. I had to go back to my church roots and knew that praising God was the only option I had. To my surprise, I was able to recruit a lot of employees to the group including people from senior management. This was a way to involve employees from every department regardless of their position at the hospital. I brought in a local pianist two times a week for our practice session, and we performed during our Christmas events which was a joy to me and a lot of the other employees. On several of our performances, I dedicated the songs to the wife of my boss because I thought this might make him treat me a little better or realize that I was still trying to be a part of the organization. Needless to say, there were people at the hospital who thought I was once again trying to play politics and gain acceptance from my boss. To be honest, they were correct in thinking that I was trying to get a little slack by using the boss' wife, but I had tried everything to fit in and nothing was working. What was I supposed to do? The next incident that happened to me that was really uncalled for was when we had our annual Christmas party for the doctors and board members. As usual, I spent a lot of time and effort to make sure that the event was well-organized and had to work countless hours with the assistance of my employees. At the day of this particular event, I arrived about two hours early to make sure that everything was set-up—the food, drinks and decorations. It was not unusual for me to pick up a broom and sweep although it was not my duty to do so. Although I never saw any other senior management person doing what I had to do, I knew I would be blamed if, for some reason, things did not turn out the way they should. I was always hands-on and although I had my staff with me, I always worked as a team. During this particular event, since I was at the event location early, I decided to have my husband stand at the entrance to welcome the guests. I felt it would be a nice gesture to greet the guests as they come in and make them feel welcome. As soon as my boss walked in, he let me know that it was not my place to welcome the guests although I was the event planner. Basically, the majority of the guests were doctors and board members and community leaders who were all White and it was not my place greet. In order words, I belonged behind closed doors.

The problem with the doctors and administration was getting worse and I found myself playing the role of investigator for my boss. We had been losing a lot of doctors, and our revenue was getting to a point where the hospital was losing money each month. We had lost a few older doctors who had been with the hospital for a while and were in key positions such as specialty areas that generated the bulk of the revenue for the hospital. I knew, at this point in time, my boss was in trouble but never thought he would be fired from his position.

The night that the board was meeting for the renewal of his contract, my boss personally asked me to stay around for the meeting in case he needed me. I was not sure why he thought he needed me as this point, but I already had firsthand information that a lot of the doctors were unhappy and wanted him fired, but the decision would have to come from the board members. As the meeting went on, it was about ten o'clock at night, and I could not stay any longer, so I decided to drive the forty-five-minute trip I made each day home.

When I got home around 11:00 p.m., I received a call from one of my peers indicating that my boss had been fired. This was the biggest shock to me because as much as I felt he had not treated me fairly, I knew he was the best person for the job. He had kept the hospital viable for several years as well as just renovated and expanded the hospital's services. He had also brought in more physicians, but his control had taken a toll on him. As I slept through the night, I did not know what to expect or think of who would be the interim administrator when I made it to work the next day.

On my way to work, I was praying as hard as I could for the news to be wrong because it is better to deal with the known instead of the unknown. When I made it to work, I found out that my peer had been made the interim administrator and it was indeed true that my boss had been fired because he refused to take a cut in pay despite the fact that the hospital was losing money and the doctors were not happy with his administration. Although the entire hospital family was in shock, I could sense that a lot of the employees were happy. It was brought to my attention that nobody ever felt this man would be fired in their lifetime working at the hospital because he was so powerful and practically acted as if he owned the hospital.

After digesting what had just happened and understanding that my boss had been fired, for a brief moment, I felt as if things were going to change for a lot of people. I started thinking of how good life was going to be for me and a lot of other employees. I knew the interim was one of us and he would immediately do the right thing. To confirm my belief, the first meeting we had was positive, and the interim immediately made a speech outlining his goals for the hospital and indicated that the hospital was for the people and he did not want to be addressed as "Mister," but by his first name. What a change from his predecessor who wanted to be addressed as "Mister."

A few days later after the firing, my old boss came to my office to speak with me. It was definitely a surprise because I cannot tell you how many times he had been in my office since I started working at the hospital. I remember only about two times, once when my father died and the other time when he wanted me to spy on the doctors. He came to my office to ask for my help in sending out his resume and to tell me about what went down during the firing. He indicated that most of the doctors I indicated were on his side and sold him out. On one hand, I felt very sorry for him and was willing to help as much as I could. I found out the hospital had given him a very lucrative severance pay and had retained a recruiter to find him a job. I was also very surprised to see the anger he had toward his firing because in my opinion, I don't think he ever thought he could be fired. As a matter of fact, he had indicated to me and others that he would retire in the community, and I am sure he thought he had played the game to his advantage. God is the only one who is sure of our future, and in this case, he proved me right. As the days and weeks went by, my former boss started frequenting the hospital a lot and asking his predecessors to help him find a job. This was not taken well by a particular senior management staff who indicated his life had been made very difficult as well. I could not understand why this individual was so upset about helping his old boss when he had been one of the persons who reaped most of the benefits in terms of responsibilities and compensation. Although I knew I did not have to speak or help my old boss after his firing, I still wanted to do the right thing by helping him acquire a job, and as a matter of fact, I still had respect for him and knew that revenge was not mine but God's. After a few weeks, I received a message from senior management that I should not allow my old boss to my office anymore and also that the hospital had paid for an outplacement recruiter to find him a job and there was no need for him to come back to the hospital. I called him and informed him of what had been said to me, and we still stayed in touch almost every day about how badly he had been treated and why I should find another job and get away from the organization.

With a new president and CEO, I knew my life was going to get better. I knew this because the interim was my peer and saw and knew what had happened to me since I began my employment at the hospital. I felt it was the right time for me to have a meeting with him and let him know I was ready to carry out some of the suggestions I had regarding moving the hospital's agenda ahead and felt he would give me the green light to go ahead. The first thing on my agenda was to ask him to rectify my salary and also let me run my departments the way I felt it should be run. To my surprise, he told me he could not do anything about my salary, and as a matter of fact, surprised me by not saying anything at all. I immediately took my case to a female board member I knew would listen and informed her that I felt my salary should be increased. She informed me that the hospital was losing money and that she could not do anything about my issue. At this point, I did not know what to think. I felt as if I had started all over again. I thought all along that with the changing of the guard, things will get better and felt that my old boss was the one hindering my success or compensation but did not know that the problem might have been my color. I also felt that the organization was still going through the process of regrouping and maybe the time was not right for me to start making any demands.

We had a lot of work to do, in the interim, with the celebration of the hospital's fiftieth anniversary, so I decided to do something big to prove that I was innovative and dedicated to my work and the hospital. As I drove home that day, I thought about writing a script about the hospital from its inception using our employees. My goal was to use the talent we had at the hospital and not use an outside organization to produce my movie. Before I made it home, which was about a total of one hour, I had come up with a rough script and how I was going to write and produce this documentary of the last fifty years of the hospital.

The next morning, I went to the information technology direc-tor and explained to him what I had in mind and asked him if he could assist me with the production. I went back to my office and had my director of marketing and the other staff and starting working on a script. I consulted with a well-known historian of the community of the history of the hospital. Within a month, I had compiled a history of the hospital and started filming the documentary. I did not mention this to my boss at the time because I did not want any distraction and I did not think he would oppose what I was trying to do. Remember, the old boss who did not want me to do anything had gone, and I was confident I had the authority to do whatever I felt was good for the hospital. Besides, I was independently producing the entire documentary so the cost was going to be very minimal. Also I was using hospital staff and making them part of the production and history of the hospital. I was amazed to find out that only fifty years ago, there was a colored hospital and Blacks as early as fifty years were not allowed to the hospital when it was built. There was a separate hospital for Blacks, and when the Blacks came to the newly built hospital, they had a separate entrance. I also found out that during my research that the Black hospital was ran by a few women who were not doctors and just one or two doctors gave their time and money to see these patients. I found out from one of the doctors who was on staff that agreed to be on the documentary that his grandmother had started the colored hospital. He was also one of the doctors that was willing to see the colored patients. I must reiterate the fact this was just fifty years ago that Blacks still had to come through separate door to get into the hospital for treatment. Although it started making sense to me as to how some of the older Black patients were behaving, I still could not understand how people could look at the mirror and go about their daily lives when they knew what was going on behind closed doors. As a matter of fact, some of the older Black patients still felt they were back fifty years ago when they visited some of the older physicians. You see, the community was still living fifty years ago, and the world did not know about it. Even with the thirty-ninth president a few blocks away, they did not care. As a matter of fact, there were a lot of the older White folks that did not care about the president because they felt he was just too liberal and paid attention to the Black plight. Though I felt sick to my stomach about the history, I still wanted to continue with the project. I looked at the history as it was in the past but still found it difficult to swallow what had happened in the past, given the fact that I could have been one of these people who had to come into the hospital through a separate entrance and wait the entire day for basic medical treatment. It started making sense to me because I started realizing that this was still happening while the world was preaching equality and in an area of the country considered a Bible Belt. One question I considered was why. Who are these people and how do they sleep at night? Did they think it was okay to treat human beings in such an inhuman way or did they really believe they were doing the right thing. Also why did the Black people have to settle for this? With all the attorneys we had around, why could someone ask for help? Is it because most of the people thought it was okay to treat Blacks in such a horrible manner? What were the preachers saying in church on Sunday? Did the Blacks allow this to happen? These were the questions I asked myself but had to move on to something else because I was fighting for me and did not have time to solve the problems of others. To be honest, coming from a different culture (Africa), I still did not understand what the Blacks had gone through during slavery although I had in-laws that were from the South and had extensive knowledge of what their parents went through. It is true that you never can feel the pain until you experience it. After all, I had the title of vice president and was being compensated at a rate that most Blacks in the South never dreamt of having. I sat in the boardrooms with Whites, I went to all the parties given for the Whites, and had to feel special according to a lot of people who heard my story. I also had a problem with the fact that with the new hospital addition, I was not given a space in the administrative suites. Each senior management staff had a significantly large office and expensive furniture. Each time I asked about where my office was supposed to be, I was ignored. I was excluded from all of the meetings regarding the new construction although I was the chief marketing officer. In the final analysis, I was given a small office in the Public Relations Department and did not even have a bathroom in my area. I had to use the public bathroom with other employees while the administrative suite had three bathrooms. The boss had a suite bigger than most and had a private bathroom and a shower. I could not understand why he decided to build a house for himself when, financially, the hospital was losing. I remember, when I brought in a local senator, he commented that the CEO's office was bigger than the office of the president of the United States. There was no reason for me fighting a losing battle.

As I finished with my documentary in preparation of the fiftieth year anniversary, I was particularly concerned with the finished product because in the past, I did not have the authority to independently come up with an idea or finish an assignment without approval from the boss or for that matter, people in lower position than myself. For example, I had a great opportunity to work with the Carter Institute to promote caregiving with the Rosalyn Carter Institute. When I brought up the idea to my boss, it was immediately discouraged. I wanted to promote caregiving because Mrs. Carter was an instrumental part of the organization, and as a former first lady, I knew her involvement would give the hospital a needed reputation and bring some form of publicity to our hospital. Also, we had an individual from the community who was walking a thousand miles to Chicago to bring awareness to mental health. I felt the hospital should participate in this cause, thereby bringing publicity again to the hospital, but to my surprise, it was rejected.

Needless to say, my goal for this particular event was to put on a spectacular event and celebrate the present and not the past. Once again, in my usual theme, I went throughout the thirty-five departments of the hospital and shot a video of each department wishing the hospital a happy fiftieth anniversary. As I finished the guest list, which included the former attorney general of the United States, the local mayor, senators and community leaders, I was very confident that the event would turn out better than any I had done before. I had tents throughout the hospital grounds, with a few television sets for viewing of my video. The entire outside of the hospital was decorated with ribbons, flowers, and white chairs for our guests. It was a beautiful Sunday afternoon, and as the crowd gathered, I felt a sense of relief and pride for the hard work I had put into the event. The event turned out to be one of the most beautiful event I had done in my professional career.

It was back to business as usual, and the position of CEO was still up in the air. After two months, the interim was named CEO, and I felt it was a good idea because someone on the senior management had assumed the responsibility, and I felt it was better to have someone who knew the organization instead of bringing in an outsider. It was time to start a new regime and the first business at hand was to get rid of the past mentality of control and make the hospital a friendly environment. You could sense the closeness of the employees and the ease in which patients, doctors, and employees walked around the hospital. There were still many employees who were loyal to the old boss that were still in shock and did not know if they would retain their jobs. For the most part, the majority of the employees were happy, and in reality, were taking advantage of the situation. The interim administration felt as if they had to prove to the employees that things would be different and made an effort to condemn the past boss.

As time went on, it seemed to me that our focus was not on the viability of the hospital but on changing attitudes. This was essential because, truly, the entire past environment was one of a dictatorship regime where almost everyone including doctors and board members were controlled by the president of the hospital. I, for one, thought the president of the hospital was just doing his job. The new direction for the hospital was one of change which was probably the most difficult job for anyone to assume. The primary goal, in my opinion, was to take care of our patients and provide adequate care for our patients and continue with our hospital goal, which was to provide adequate care for the patients in the ten-county area. Although the hospital had named a permanent CEO, and as much as I admired my new boss, I knew he was going to have a difficult time getting the organization in the direction I felt it should be. At this point in time, we had several factions that were battling one another for power, and the business of taking care of patients was not being taken care of. My new boss, who was a soft-spoken individual and had been at the organization for quite a while, felt he had to concentrate on winning the respect of the employees and reassure the community that the hospital's reputation would change from what it had been in the past. In all reality, this situation was a difficult one because the organization had been losing revenue and the doctors were leaving. Remember, it was a known fact that the past administrator controlled the hospital, and in order to change the perception, the new administrator felt it was his duty to change the perception. For a period of time, we started implementing ideas geared toward the employees; for example, rewarding employees for good customer service performance. Each month, an employee was recognized for his or her contribution to the patients or to another employee. For a while, this process went on okay, but it became a competition among the senior management staff. It also became a popularity contest whereby certain people were appointed to the committee and were instructed to do as they were told. Most of these individuals were middle management staff who felt they could do anything without any consequences. As indicated earlier in the book, it was a known fact that most of the Blacks were working in the kitchen or housekeeping and did not have the authority to even speak, let alone be put on any committee. So as you can imagine, only one or two Blacks were represented in a position that had any say-so in whatever was going on at the hospital. Even as an executive, I was limited to an extent to what I could do. The majority of the employees respected me as a vice president, but I knew I had to ask for approval before I carried out any plans. One particular instance was when I needed furniture for my office, which I had my secretary order from the purchasing department. I waited for about three months and was told the order was canceled because the middle management individual who was supposedly ordering the furniture did not. When I asked why not, I was told I did not need furniture and should be glad to use what I had already, although everyone from senior management to middle management had new office furniture. I could not understand why certain members of the middle management team had more authority than me, but as time went on, I realized that it was not my position but my color. I was even told that it was a known fact that all of my orders had to go through administration regardless of how small or big it was. For the life of me, I could not understand why I was being treated this way and the rest of the executives did whatever they wanted. As a matter of fact, I had to get one of my peers to approve any of my requests and also had to make treats before anything was done. Because of these situations, a lot of the staff did not give me the respect due, and as a matter of fact, treated me as if we had the same position, and in some instances, acted as if I worked for them.

Not knowing what to expect from my new boss, I continued my daily routine but noticed most of my responsibilities were being handled by consultants. Although this process did not make sense to me, I understood where he was going from because the hospital was at a point where we were barely surviving financially. My objection of using consultants was because these consultants came to me for information about the hospital and then turned around and submitted my ideas to the boss. Also, with the financial situation of the hospital, I could not understand why we would pay a consultant to come in and get ideas and information from the employees. It got so bad that we had consultants working with almost every department of the hospital. I started questioning the fee for my consultant because the bill was coming to me for approval and I could not understand why we were paying for basic letter writing for my boss when I was more than capable of writing, and for that matter, I had an excellent writer in my department. For the last several years, we had produced and written numerous publications for the hospital as well as produced a monthly newsletter for the entire organization. I knew in order to keep my job, I had to be friends with the consultants and make sure they got paid.

As things started settling down, I realized that the new boss did not fully take charge of the organization. He named one of his best buddies senior vice president and turned over the daily administrative duties to him. I was kind of happy that he appointed this individual because in my opinion, he was committed and knew the inside and out of the hospital. The one problem I think he had was relating to employees. He could not look you straight in the eye, but to his credit, was one of the most hard working individuals I had worked with and knew the hospital business. He was a dear friend to me, and I could talk to him about a lot of personal and work-related issues without worrying that he would sell me out. As much as I respected him, he was one of those people who would not speak up for anyone. After his best friend took over as hospital CEO and president, I felt he now could advise him and direct him on dealing with people and giving me the due respect I deserve. At this point of the game, he was not a punching bag as he had been with the previous administrator. At one point during the previous administration, he was the one that did the dirty job for the CEO. One instance was when I received a letter from the CEO indicating I had submitted more travel mileage on a trip we both took together to a conference and the fact that he did not support me was beyond my imagination. When I confronted him why he did not support me he told me he could not say anything and was not willing to jeopardize his job. There are many instances when he could have supported me but chose not to. I, on the other hand, understood where he was coming from and knew he had no choice but to do what he was told to. He, like everybody, also had skeletons in his closet so he could not take the chance to support me without serious repercussions.

Even with the new CEO in place, I realized that I was back at the same place I was a few years ago. I still could not understand why I was not being included in decision-making and why an individual that had worked with me for the past several years would ignore my contribution to the hospital as a whole. I immediately noticed—although his characteristics were far better than the previous boss—the same pattern in terms of my ability to perform my duties, and my understanding was that he was dealing with the pressure of assuming a new role and had to give him more time to settle in and tackle the numerous problems we had facing the hospital. I waited and waited but things got even worse. I noticed that I was being left out of committees where senior management was involved. One instance for sure was when a strategic committee was assembled, and each member of senior management was included but myself. When I found out about this particular committee, I was furious and wanted to know why I was not included in a committee which I felt I should have been the chair since marketing was an integral part of the committee and I was the chief marketing officer of the hospital. The CEO refused to return my calls on this particular issue so as my husband has always told me to follow up any inquiry in writing, that is what I did. As soon as my e-mail was received, I received a response by phone from the administrative secretary indicating I had been included on the committee. This was the beginning of trouble for me because I noticed that I was being set up and asked to make presentations on issues without any notice for me to gather pertinent information to assist me in my presentation.

On one night, at a board meeting out of the blue, my boss asked me to make a presentation on a subject that I was not prepared to talk about. Also, I was still very unhappy with the fact that I had not been included in a lot of what was going on as a senior management team member., so as I got up to speak, the only thing that came to mind was anger and I could not even speak. My voice was shaking, my blood pressure was so high I could hear my heart beating, and it felt as if the I was going to pass out from the heat although it was a cool evening. I obviously made a fool of myself because every word that came out of my mouth did not make sense. I could see the people in the room looking at me as if I was dumb and stupid, and I also could feel as if they were thinking, *How did this woman get her position as an executive?*

As the meeting ended, I immediately walked to my boss' office and told him flat out right that I was embarrassed and felt he caught me off guard and that I was very upset before the meeting for not including me in some of the very important issues of the hospital. I told him that I felt he did me wrong and that I wanted any chance to do a presentation after I did some research on the subject matter.

I should make it clear that when I went to his office, I was prepared to be fired because I was not holding back how I felt and also I wanted him to know that I had been told by one of the doctors that he deliberately left me off the committee. I also, in an a roundabout way, told him he asked me to speak that night to make me look bad, so that he could justify treating me as someone who was incapable of doing their job. This game was not happening because I had been a victim of it for quite a while and knew exactly how to handle myself. I had done more than enough and had a favorable reputation at the hospital including winning numerous awards from several hospital associations to feel comfortable about my abilities to perform.

Even after our conversation and another shot at continuing to do my assigned duties, I realized that I was back again at dealing with my boss' wife. I felt that getting her involved in some of the events at the hospital would be good for me in terms of building my relationship with the boss. I contacted her and asked her if she would help me with the hospital's annual party as well as start visiting our doctors.

She was receptive to my ideas, and we started on a good foot, visiting the doctors and planning several events for the hospital. Although I enjoyed working with her, I realized that she did not know how to handle the doctors. She made several remarks to some of the doctors that I felt were inappropriate. It was my job to smooth over what she said to those doctors and move on. After a while, I found out that she was beginning to want to take over my job. When I tried to resist, the next thing I knew, my events, especially the Christmas Party for the doctors and employees, had been taken away from me, and she was working with another employee from a different department who was not part of my team. I went to the one person who always listened to me to ask why this was happening and was told to keep my mouth shut. Obviously, I could not go to her husband to complain about his wife. As a matter of fact, it was known around the hospital that she wore the pants in the family.

After a while, the stress from the several years of dealing with my old and new boss started taking a toll on me. I became very short-tempered with my kids and husband and found myself spending the majority of my time in my bedroom. I was so tired each time I came back from work that I stopped attending any of my children's school activities. I could not get my husband to understand what I was going through because as much as he tried to advise me on what to do, I could not afford to listen to his advice and lose my job. We needed my salary to keep the children in private school and live the life I thought we had worked hard for. In all honesty, my husband asked me on several occasion to quit my job and felt I was more than qualified to find another position elsewhere. He was so fed up with my daily complaints that he basically asked me to sue the hospital or stop complaining. My health started getting bad, and all I could do was pray as I got to work for a calm day without the pressure of dealing with people around me. I must say in order to continue with the pressure, I developed several personalities. I would get to work and act as if everything was okay one day; the next day, I would get to work and bring out another personality which was to fight anyone who confronted me. Because of all of these circumstances, I started developing panic attacks and could not stand going to work. On one trip for a conference, as I was approaching a bridge which I had gone through this bridge for several years. I suffered from chills and could not breathe. I immediately pulled over and had to sit on the side of the road for almost ten minutes before driving again to the conference. This was one of the times I felt the job was not worth it, but what choice did I have? I had a family to take care of and could not bear the thought of not providing for my children.

As soon as I got to the conference, I met with one of the doctors on staff to tell him of my experience and told him I had just suffered a panic attack. He informed me to come to his office as soon as we got back home so he could prescribe me medication for my condition. He thought my condition was serious so he made sure he drove behind me on my way home after the conference to make sure I did not have another panic attack. As a matter of fact, he felt my condition was serious enough for us to re-route our trip in order to avoid traveling over a bridge or areas that might trigger another panic attack.

A few weeks after the conference, I thought my condition was going to change since I was not working under pressure, but it did not. I immediately knew I had to consult my doctor who decided to put me on anti-anxiety medication to prevent my panic attack. This was the first time I had been on any other medication except my blood pressure medication which was a result of the stress I had to endure with my job and most of all dealing with situations that basically could have been prevented. I had to keep this illness secret because I did not want my superior to have any reason to try to get rid of me or insist I was not capable of functioning. Although I had medical insurance to cover my visits to the doctor as well as pay for my medication, I was forced to pay out of pocket. I also made sure I went to my consultation when the office was closed so I did not run into patients, or for that matter, anyone from the organization. Although a lot of patient information was supposedly confidential, it was a common practice to discuss these issues openly in meetings, and frequently without the concern of the consequences, because certain people in the organization assumed they were immune to prosecution and were literally above the law. It was not unusual because, as mentioned earlier in the book, most of the patients who sought patient care from the hospital were of the mindset—in my opinion and experience in dealing with them—that they could not challenge whatever was said or done to them. It was seldom that a lawsuit was brought against the hospital by a poor person because most of the problems or issues that would have helped their cases mysteriously went missing, be destroyed, or tampered with. I cannot attest to the fact that this was done to everyone or to help the doctors, but I can say that the powers that be felt the hospital was above the law. I must say that in the event of an occasional lawsuit, it was seldom that the organization settled a case even when it was clear it was their fault. As a matter of fact, the lawyer representing the hospital had his son's tuition for medical school paid by the hospital although he was a very rich man. I could not understand why the hospital paid for his son and could not pay for kids that needed the assistance of the hospital.

Business at the hospital continued as usual, with the hospital losing a lot of money and doctors leaving because they could not make ends meet and the present administration did not seem to have a good plan for growth for the hospital. The working environment was not getting any better and although we had a new CEO, in my opinion he did not have control of the organization. He spent more time trying to defend his position as CEO as well as try to change the culture of the organization which had been one of a dictatorship. To be fair, he tried to make everyone comfortable and relaxed at the work place, but the problem was nothing was changing in terms of how certain people were treated. The Blacks were still not given the opportunities the Whites had in terms of compensation or given advancement in the employment.

As much as I wanted to help the rest of my people, I had to look out for myself. Things had not changed for me either, and I got even more frustrated because I hoped for changes and thought because the new COE was once my peer and knew how I had been treated in the past, he would make sure I was given the respect due. For a while, I thought it was the pressure of assuming a new position and changing the culture we had, but as time went on, I noticed things were the same. It was more difficult for me to deal with the situation because the same treatment that was done in the past continued. The Black employees, although could walk around the hospital freely, were still confined to the laundry room or kitchen which was what was felt was the right position for them. In other words, nothing had changed from the past administration, and no one, even the board members, cared enough to make thing right. All the Blacks wanted was a little respect or treated as if they part of the organization. At my boiling point, I started making little comments about without the kitchen help or laundry help or the nurses aid, the hospital would not be where it was. It was true, in a sense, because most of the minimal jobs were done by the Blacks, and as low as they were compensated, they continued to do their jobs. This was the same with our doctors in the sense that most of the doctors on staff were Whites and could not do anything wrong. They had the right to talk to patients as if they were not human beings and were never reprimanded. Each time the White doctors broke some of the hospital rules—which could be not maintaining his or her medical records or abusing a hospital worker or patient—it was considered a misunderstanding and the complaint was sent to the ethics committee which obviously would be dismissed. On the other hand, if a Black doctor had a bad day or did anything out of the ordinary, it was not unusual to suspend their privileges. There was a female doctor that was always at the hospital each morning I came to work that actually cared for her patients and passionate about her work to the community as well as the patients. She always participated in hospital activities as well as treated each individual regardless of their position at the hospital with respect and dignity, yet, in my opinion, she was treated by administration and some of the older White doctors as dirt. She was criticized for speaking up for what she believed in and not taking crap from anyone. She was one of the doctors that would pay her bill on time while the other White doctors that owed the hospital money basically did not have to pay or did pay when they wanted to.

The usual routine of running the hospital comprised of meetings which, 90 percent of the time, comprised of middle and senior management. In these meetings, approximately 2 percent of the people in attendance were Blacks and the rest Whites. In our senior management, I was the only Black person, and in the beginning, you could see the tension in the room because I knew I was not wanted in the meetings. This was, in my opinion, because they did not want me to hear what was going on behind closed doors. At most of the meetings, a disclaimer was always made indicating the meeting was confidential and nothing said in the meeting should be repeated. I was no fool and knew they were directing these comments to me. I knew that information discussed in the meetings were to be kept confidential regardless of how I felt, but for some reason, each and every time I was present at our executive meeting, the point was always made to keep everything confidential, which made me feel it was directed to me. It was almost like a threat to me and some form of intimidation. The point to keep everything confidential was not something I thought should be mentioned because all patient information or any form of information was supposed to be kept confidential in the first place. The reason for this proclamation at most of the meetings was to make sure I did not say how certain individuals were treated. It was obvious because some individuals had preferential treatment, and they could do whatever they wanted to without repercussion. This was very evident with one White doctor who had been involved in what I would call uncalled-for behavior and gotten away with it each time. This was not the same with Black doctors or employees. A good example happened when one of our Black doctors was accused of not taking care of a patient although she was the primary caregiver and had referred the patient to a consultant who was then taking care of the patient, and as a matter of fact, the incident with the patient happened after the patient had been released from the primary care physician's care. I am certain this happens with many organizations behind closed doors, but it was done in my organization without any care of repercussions or concern for the patients or the families that would be affected.

I asked myself several times why I was putting up with such an envy organization, but once again, as a Christian and based on my beliefs or upbringing, I hoped and prayed things would change. I convinced myself each day, someone would see the same thing I was seeing and bring it to the forefront. Also, I was too afraid to be the one to blow the whistle due to the fact that many people depended on their jobs to feed their families and being the second or third largest employer in the region, most of the Blacks worked at the hospital, and it would have been very difficult for them to find employment elsewhere due to the fact that most of them were rather poor and could not afford going to distance. Another factor was that some of their families had worked at the hospital for most of their lives, and it was almost a given that these people would eventually follow the same path.

Needless to say, as much as I wanted to be the one to address the issues I felt were not fair to the Blacks at the hospital, I could not do the job alone, and as a matter of fact, did not have the authority or support to make any changes. As mentioned, always hoped each day that God would come down and change everything for the best. As the saying goes, God does not give you more than you can bear and he works on his own time and you cannot push or tell him what to do but wait for the right time. I always knew, sooner or later, he would bring some form of change to all the atrocities that were going on at the hospital. As much as I do not want to believe that God was watching what was happening to me and others, on March 1, 2007, I was attending a conference about 200 miles away from the hospital and received a call from one of my physician friends informing me that a tornado had just hit the hospital. I had been watching the storm in the neighboring state the entire night and knew that the storm was moving toward the area of the hospital but did not for one minute think it would hit the hospital. Who in a million years would think a hospital that is meant to take care of people and bring them back to life would be hit by a storm or destroyed otherwise? I could not believe what I was hearing, and at the point in time, I did not even ask if the hospital had been destroyed or if anyone was hurt. I immediately started calling my staff and anyone I knew for more information. Unfortunately, most of the phone numbers I was calling were not working. Finally, I got in touch with my kids about 50 miles away from the hospital to find out if they had heard or affected by the storm. I could not get any crucial information from them since it was early in the morning and they did not have any news due to the fact that the television station they were listening did not cover that area of the storm. My children informed me it was raining hard but no apparent severe weather in their area. I called my director of hospice who was attending the conference with me and informed her of the situation at home. It was also pouring rain where we were, so we had to wait for the rain to stop before leaving the conference. I called some of the folks back at the hospital and found out that hospital personnel had been tracking the storm and were preparing for the worst as the storm was approaching. As word came that the storm was heading toward the hospital, the employees on staff that night moved all of the patients out into the hallway away from the windows. Once, the storm hit the buildings, the patients were immediately taken from the hospital with the assistance of city buses and local emergency officials. The storm hit at 9:26 p.m. which was classified as an F3 tornado that packed winds upwards of 160 miles per hour and left a wide path of destruction throughout the city, and the hospital was the heart of the destruction.

It was only by the grace of God that all fifty-three patients were evacuated and no lives were lost. Everyone in the community came to the aid of the hospital, including, and not limited to, the physicians, local officials, and volunteers, young and old. Although they were two lives lost in the county, it was evident that the people from this small county were there to face the reality of the storm.

After hearing the impact and the condition of the hospital as well as the county, I knew I had to get my bags packed and ready to make the trip back home. I was in charge of public relations for the hospital, I felt it was my duty to take a chance on the weather and start heading back to the hospital.

After we checked ourselves out from the hotel and got into our car, I was so nervous that I had to have my director who, at the point in time, was much older than I drive us back home. As we drove back home, I could only speculate about the condition of the hospital and the city, for that matter. I did not know what to expect because I had never been in a tornado or experienced the impact of a tornado. As we got to about two miles to the hospital location, we noticed the road leading to the hospital had been blocked and as a matter of fact, saw a town that had been completely destroyed. I did not even know where I was because buildings were down, power lines hanging, and debris all over the place. As we made our way closer to the hospital, we were stopped and asked not to enter the zone because it was too dangerous. I immediately informed the officer on duty that I was an executive from the hospital and had to get to the hospital as soon as possible. I also had to show them my hospital badge in order for them to let us through. The officer informed me I was entering that area on my own risk. At this point in time, I was not thinking about myself but how to get to the hospital and offer my help.

Usually a hospital is the place that people turn to for care, and hopefully, a safe place, but it became a place of disaster, leaving behind a trail of broken glasses throughout the entire hospital, tossed cars, twisted cars, fallen power lines, and debris. I could not believe what I was seeing with my own eyes because it was just a few days I had been walking the halls and tending to business as usual. It was also amazing to see what could happen in a flash. It is so true when you hear people say you could be here one day and gone the next day, or how powerful God's wrath can be.

After a few seconds or minutes of daydreaming and realizing that the hospital was in total destruction, for a split moment, I had to compose myself and think of what or how my life was going to be affected by the destruction. All I could imagine at the time was my children. How would I provide for them? What is going to happen to the new lifestyle we had been living? All these questions were popping in my head, and for a minute, I saw everything I had worked for disappear into thin air, figuratively speaking.

I came back to reality and knew there was no time for daydreaming or feeling sorry for myself. I made my way through the debris to find out where my boss and peers were to see what had to be done. When I found the rest of the administrative team, the looks on their faces told the same story. It was truly a sad day for all of us and the rest of the employees. As we met to discuss the fate of the hospital and the employees, it was apparent that we had to immediately decide where to see patients the next day. Our community could not be without a healthcare organization because most of our patients could not afford to travel outside of the immediate surroundings, so it was imperative we set up a system whereby we could accommodate our patients. The executive team met later in the day at a board member's house to strategize on a plan or plans to deal with the tremendous burden we were facing. Some of us sat on the floor and the others by the table and threw out ideas on what to do. Some of the thoughts were good, and some were, in my opinion, unrealistic, but this was not the time and place to argue or fight about who was right or wrong. It was immediately agreed that we set up a temporary clinic at one of the local churches to provide minimal health care for the community. There were many people who were injured who needed immediate attention who were unable to go elsewhere, so it was apparent that we had to establish a makeshift location to start the process of giving the basic care we could provide given the circumstance we were facing. We collaborated with the local Red Cross Organization in seeing the patients who needed basic care at our local Baptist Church. The Baptist Church became our command center for the community to come and get treatment or find out about their loved ones. Also, on the same day March 2, the hospital put together a press conference at the local university to inform the community what our plans were in terms of providing medical care and also to inform the employees what was going to be taking place moving forward.

As the executive for the hospital in charge of marketing and public relations, I was responsible for getting the information out to the numerous press representatives that had flooded the city. It was also necessary for us to pass on this information to our employees and their families. We had to use any and all resources needed to get this message out. We had to depend on our local internet provider to help us get the message out, obviously, because our offices had been destroyed. I was running like a chicken with her head cut off, and for some reason or the other, God gave me the strength to go on even though I was on my feet all day. There was so much coordination between many agencies that at one point, I felt overwhelmed and wished this horrible nightmare had happened to someone else.

A number of dignitaries visited over the next few days, including the president of the United States, George W. Bush; our thirty-ninth president, Jimmy Carter, who was a resident of the community; Georgia Governor, Sunny Purdue; Congressman Sanford Bishop; the president of the Georgia Hospital Association; and many more. President Bush took a helicopter ride to assess the damage and declared it a devastation.

With all this happening, I felt it was my duty to head the process of getting the information out. I started noticing that my role was getting smaller and smaller, and I was being left out from decision-making and found myself being treated just as an ordinary employee without the authority to make decisions. An example would be when President Bush was scheduled to visit the site of destruction at the hospital, and as evident, the security was going to be imposed. As much as I knew the president's visit was going to be brief, I wanted to have the opportunity of meeting our president. It is not every day that a sitting president comes to your community, so I was excited and hoping to at least have the opportunity to touch or shake the president's hands. I got up early in the morning after working twelve to sixteen hours the previous day to make the one-hour drive back to the hospital to be part of the crowd that was there to welcome the president to our community. When I arrived at the hospital, I noticed all of my peers were already at the site, and as usual, there was so much security around that I could not make my way into the area. I called one of the other vice presidents to find out where she was, and she informed me make my way to the location, but unfortunately, I was told by a Secret Service individual that I was not allowed to be in the area without permission from my boss. I immediately called my boss on his cell phone to let him know that I needed him to give permission to the Secret Service agent for me to get in. When I did not get a response from him after a few seconds, which seemed like minutes, I called my peer and asked her to let him know that I was trying to get in. She informed me she was by him and told him I was out there to give instructions for me to get it. I waited for a while and nothing happened, so I went back to the makeshift office we had and felt as if someone had hit me on my stomach. For a minute, I was so angry that I did not get to see the president and felt I had the right to see him. After all, I was an executive, and in reality, although I was not the chief executive officer of the hospital, I had the right to be there. It was my right after the position was given to me. How dare him do this to me? I was furious—I mean furious—for a minute for all the past and struggles that I had endured came back in the form of tears. I was excluded from what a lot of people would consider once-in-a-lifetime experience. As I sat in my little shack working area, I starting thinking of what I was going to say to my boss. I was going to confront him and let him know how disappointed I was and just let out my frustration on not just this particular incident but how I felt cheated and now and in the past.

After the president left, everyone, especially my peers who attended the event, came back marveling about meeting the president with pictures and conversations of how it felt. The more they talked about the event, the angrier. I became I could feel the effects on me because my head was throbbing, and I instantly knew my blood pressure was up. I could feel the tears wanting to come down my face, but I refused to cry in front of so many people. I had planned my revenge, which was call the NAACP president and tell him exactly what happened, but I immediately dismissed the thought because I knew the reputation of the president of the NAACP was not favorable in the community. I stayed around the area waiting to confront my boss, but fortunately for me, he had left the area and I did not get the opportunity to let him know how I felt. Maybe God was watching over me and wanted to protect me from being fired due to the fact that I was determined to chop his head off, literally speaking.

I drove home just thinking what it would have been to meet the sitting president of the United States of America. I consoled myself by knowing I knew another famous president of the United States and knew me by name and had many opportunities to work with him and most of all, sat next to his wife during his seventy-fifth birthday party. I also remembered that President Carter, during his seventy-fifth birthday party, greeted me first when he walked into the press room with dignitaries such as Sam Donaldson, Spencer

Christian, the Dixie Chicks, and many more. As far as I was concerned, they could not take that experience from me. As much as I tried to justify the incident, I could not let it go.

The next day, I came to work and during our daily briefing, I made it a point not to say a word to my boss. He knew how I felt because everyone in the group at this point knew that I was upset. He didn't say anything to me or try to apologize or try to cover his behavior toward me. The impression I got from him was, "Screw you. Who do you think you are? You work for me and I can do what I want regardless of your feelings." In a way, he was right because he was the boss, and in Georgia, it is at-will employment and I could be terminated at will. I had to tell myself many times that he was under a lot of pressure being the leader of our hospital. I wanted to give him the benefit of doubt, and due to the fact that there was so much going on, I had to concentrate on the task at hand.

As the news of the destruction went cold, I knew I had to get the press inside the hospital where the real destruction was. It was very difficult to get inside the hospital because it was considered unsafe after the second or third day because mold was developing on almost everything inside the hospital due to the water. I knew it was dangerous to get inside, but I also knew it was important to see the condition inside which would show the actual destruction of the hospital. I felt the outside of the building was not as horrific as the inside, and for us to get any media attention, we had to get reporters inside of the building. To my surprise, I was denied access to the inside of the building by the security director, who was considered a subordinate to me even when I knew it was where the action was. It was actually embarrassing for me because I had identified myself as the vice president of marketing and public relations for the hospital and basically was the point person for reporters. I immediately called one of the other people at my level, and he also told me I could not get into the building. At this point, I am thinking, *What the hell is going on? Can't these people see that the story is in the building?* If we need support from across the country, they must see what has happened inside the building. We had walls that had been ripped off; branches in patients' rooms, a whole side of the third floor blown out; the administrative office board room completely destroyed, with broken glass everywhere; and the entire second floor destroyed. I said to myself, "I was going in and will ask for forgiveness later."

I asked a reporter from CNN, knowing that they had seen worse or had been in more dangerous situations than what we had if they felt safe to go in. They were delighted because, as any reporter knows, they want something that tells the story. I decided to go inside the building with a few reporters, and after about forty-five minutes of going through complete darkness without any form of direction, I realized I could not tell where I was. This is a building that I walked through every day for the past ten years and was totally lost. As we made our way somewhere in the building, I heard someone shouting my name: "Comfort Green! Comfort Green, get the hell out right now!"

I ignored the first outburst until I heard my name again, but this time I knew I had to get out. The reporters that were with me told me they had to get more footage and were not going to be pulled out until they had enough. I stayed with them and made sure they had covered the entire interior of the hospital. I knew I had done my job despite the resistance from my superior because after the airing of the footage from inside the building, it was apparent the damage outside was miniscule compared to the damage inside the hospital.

The task for me and the others responsible for the hospital was to think about building a facility could accommodate and offer more services to the community. We had moved from the church to a tent, and now it was time to focus on building something more stable and safer than what we had. We were blessed to have Federal Emergency Management Administration (FEMA) and Georgia Emergency Management Administration (GEMA) working with us to plan and quickly rebuild a facility. This process meant meetings and more meetings, and at the same time, keeping a focus on what we were facing at the time. As a result of the storm, we had lost a lot of our doctors, especially sub-specialists, so our level of services were limited. We delivered our first baby after the storm ten days after the storm in a hot tent with very limited resources. It became apparent that the process of building a new and stable hospital was the number one priority, and it was estimated that it would take about three months. With the financial help from FEMA and GEMA collectively, it was estimated that a new temporary facility will be ready in three months. With the bureaucracy of any government issues, it took us quite a while to meet the goal of rebuilding in three months. We had to decide on what type of material to use for the building, and several parties involved in the planning had to agree. As usual, when more than one boss in involved in any negotiation process, it is not an easy task to accomplish. The ceremonial process in my opinion took precedence over the urgency of building a hospital which was priority. The politics was one that I could not tolerate, but I was not making the decision. As a matter of fact, although I wanted to believe that I was part of the team, in actuality, I was just a body in the room. I had the authority on paper as part of senior management but not to open my mouth. I felt I had some good ideas to share, but I was busy taking pictures or acting as a maid. None cared about the many employees who had lost their jobs although the hospital had business interruption insurance which meant most of the employees were paid. I felt that this was not a good idea to pay folks who were not working when we were trying as hard to generate funds to build a new hospital. The fact was the government had declared the area destroyed so there were federal monies to rebuild. In retrospect, this is the reason why this country is going bankrupt because we depend too much on what we think is free money. We had over 300 employees who could have brought themselves to work and save the federal government from wasting much funds for when it truly needed.

After the several weeks or hashing out the details on how and who was going to pay for the rebuilding of the hospital, it was agreed that we get units called OGIM units to start the process of building. We had a ceremony which was headed by FEMA and the other organizations involved. This was a learning process for my department because I had to put together the entire ceremony. It was quite an experience because I had to work with the best qualified individuals from FEMA. Once again, although I had direction from an elite organization like FEMA, my boss still felt we needed help from our consultants. Once again, we were getting help but had to pay more money out to an outside organization because it was felt that I could not do my job. As much as I did not want to work with the consultants—because it was a waste of money and the consultants were helpful but they basically got all the information from me and then gave the information back to me—it does not make sense, but it is what happened. On several occasions, I had to question the charges that were submitted for public relations and marketing consulting because it was outrageous and I could not consciously let it keep going on. When I could not get to my boss about these charges, I turned to another peer that was truly concerned about the finances of the hospital to kind of get his input about the situation. He agreed with me that the charges were too much and still could not do anything about it. The point I am making is that there was no need to hire a consultant for a job I felt I could handle just as well and had been doing so. I am not claiming to know everything because we could always learn from a situation or from each other. It was obvious that no matter what I did, it was never good enough. I found myself asking why, for a long period of time, I was the only senior manager who had an advance degree from an accredited university yet not given the chance or basic respect to do my job. I wanted to at least fail or fall a few times before just being flatly rejected. The only one answer I had was it was a matter of me being a woman, and a Black woman, for that matter—"*Still colored.*"

Looking forward, I had to concentrate on what was important for the community and not on what my problems. The rebuilding of the hospital was the main focus for everyone, and that was what we had to concentrate on. With the assistance of my team, I came up with a campaign to calm down the rumor that the hospital will be closing. To cool down the perception and rumors in the community, the "Indestructible" campaign was created to let the community know that the old building was destroyed but the commitment to build another hospital was still there. The "Indestructible" campaign was a big success and had a tremendous impact, even on a ten-year-old boy who raised one thousand dollars to save his hospital. Many other individuals, after realizing that the hospital was a complete loss and that they had to get involved in whatever way they could to bring back so form of health care to the community as quickly as possible, starting getting involved in their own fundraising. Volunteers started calling my department, wanting to participate in activities that would raise money for the hospital. I saw the progress of getting a better place where patients could be seen improved.

Within two months of the disaster, FEMA trailers were set up to provide a temporary facility housing twenty-four-hour urgent care and outpatient services for the community. This made it possible for the hospital to provide limited amount of specialty services, such as labs and X-ray, which prevented local residents of traveling out of the area. This new setup was a major improvement, because the day after the tornado, we saw patients at a local church, then transitioned to tents, and the big move into a trailer. Over a period of less than two months, we had seen about five thousand patients. To keep the focus on the hospital, the Mayor of the city declared the hospital "Indestructible Day," which gave us more leverage to raise funds to put toward the rebuilding of the hospital.

I also organized a fundraiser with one of our local NFL players who grew up in the community. He was able to bring a number of his NFL friends to the community for a weekend of activities and raised quite a bit of money. Another big fundraiser was a contest sponsored by Siemens between 100 hospitals over all the United States to win an MRI machine. This contest was for a state-of-the-art MRI machine which cost over $800,000. Although we were faced with the destruction, the entire employee base and community stepped up, and my marketing efforts, which included using e-mails, yard signs, and countless other techniques were used to get votes, and when it was all said and done, our hospital had more than 260,000 votes which was 100,000 more votes than the nearest competitor. Although we did not win the contest due to the fact that we lacked a facility to house the equipment, the MRI company officials were impressed with our efforts and donated a brand-new machine when the hospital moved into its permanent facility. Once again, this brought attention to our community and exposed the need for more help from not only those around us but from media folks that had worked with Katie Couric. I had never worked so hard in my life and have never been a part of such a wonderful experience. Here I was, working with top-notch producers and being seen all over the country. With this achievement, I felt everyone was going to respect me and give me the credit that was due to me. I was always looking for that acknowledgment even when I knew it was not going to happen. I did not want to believe that the entire world was seeing my achievement and yet those around me could not give me the slightest respect for doing a job they were not capable of doing. What was it going to take for anyone to say, "Yes, yes! Finally made it." I am the type of person that gives everyone a chance regardless of their ability because I believe that everyone created by God has something to offer. On the other hand, I am also someone who likes to be recognized and given credit for my success regardless of how big or small. With all these numerous successes especially the winning the MRI machine, I was totally surprised that the credit was given to someone else who worked in a different department of the hospital and was not included in the team until after the process of putting a video for the contest had started. As a matter of fact, we did not officially win the contest because my boss had our consultants get involved in the process which was against the contest rules.

As plans continued for the rebuilding of the hospital, I focused on doing what I thought was the right thing to do, which was try as hard as possible to raise money for the hospital. I came up with the idea of selling the old bricks from the hospital because the hospital had been an institute to the community. I presented my idea to the boss, and he gave me the green light to move forward with the project. I personally went to the site of the destruction with my staff and picked the bricks for days which were engraved with the name of the hospital and the date it was destroyed. I had a local artist engrave the bricks, which were boxed and sold in different prize range ($25, $50, and $100, respectively). This was a success because many people who either delivered their babies or were born at the hospital wanted a souvenir. All these ideas were thought of by me alone, and although I had to persuade my boss for approval, most of the time, I went ahead and implemented the ideas without approval. No one cared about what I was doing or if I had any input on the rebuilding of the hospital. It was common not to include me in the decision-making or planning of the rebuilding of the new hospital. I attended some of the planning session as an onlooker and took pictures during the meetings or called in lunches for the participants. My involvement in any decision-making was none. As much as I felt left out, there was enough for me to do with and for the employees. I felt this was an issue of my race and gender, but the other female executive on the team had access to the meetings. The answer I was given each time I asked why my name was not included in the planning was: "Nothing." I was ignored, and although it was obvious to the secretaries at the administrative offices that I was always left out, they did their best to find a good excuse to keep me motivated and focused on what I was doing. I was constantly getting rave reviews from the community about my participation and updates of what was happening with the construction of the new facility. People were coming to me and basically telling me stories of how fortunate it was to have a hospital and how they never dreamed this would happen to them. On the other hand, some were also complaining that it was taking too long for the hospital to start the process of rebuilding. In my opinion, we were not giving enough information to the community and some higher-ups felt it was not necessary to tell the people we were serving what was going on. As I realize now, we were conducting business as if we did not have to keep the citizens of our serving areas involved. The only answer I have why it happened is the fact that we were serving a lot of poor and underprivileged individuals who had no money, could not travel out of the county, and could not articulate how they felt about the hospital. Most of these people looked like me, had the same ethnic composition like me, and probably did not think or had been told all their lives they were not worthy of anything. It was my responsibility to make sure things were moving in the right direction even when I had to fight every day to maintain my sanity. It was not about me anymore. It was for my people and those who did not have a say. I had the title, and that's all I needed to move forward.

The new facility a seventy-six-bed, 71,000-square-foot facility was opened a year later. Even with the collaboration of FEMA and

GEMA, it still took us over a year to build a facility that provided critical services such as surgery, critical care emergency room, pediatrics, and in-patient rehabilitative units because of the politics amongst the groups. FEMA thought we would complete the building in three months, but as mentioned, it took over a year to put up a temporary facility. As much as I had very little input to the rebuilding of the hospital, it was quite a difficult experience for everyone involved. I saw many of my peers age over a period of just a year. I am not sure how we made it through this initial period, but now I can say it was by the grace of God. As the mayor said, "They say that this is the first time that anything like this has been, and I cannot think of any better community to have it. This just symbolizes the commitment that this community has to our hospital and to our community." What the mayor was referring to in his speech was that the material used to build the temporary hospital was one that had never been used before. This was the idea of FEMA to use this fabricated material already assembled to build the hospital. The idea of using this type of building material was to expedite the process. As one would imagine, whenever you are dealing with any form of government process, it takes a long time regardless.

The opening of the new facility was a grand celebration for all parties involved in the building as well as the community. My department, with the help of our consultant, FEMA, GEMA put together a spectacular show for the community. I was fortunate to deal with an experienced public relations individual who had worked with the best to help me navigate the program. I spent countless hours with them to put together the program and make sure everyone was involved. This was my moment to shine and take control of the program and the entire event. The only reason I was given this opportunity to head the event was because the folks at FEMA thought with my position, I was the one to go to. Little did they know that if the powers above knew that I was in charge, they would have stopped the process. The saving grace for me was that my boss thought I was working with the consultants and did not make any decisions. On the other hand, the consultants were happy to let me be the spokesperson because I was the person approving their invoices. Although I took a lot of abuses from this particular consultant, I had no choice but to work for them. Every time I received a call from them—which was almost every day—my first instinct was to be uncooperative. I must also say the owner of the consulting firm was always gracious to me and felt that I was capable of running the show. I consulted him when I was hiring a director of marketing and had to let him convince my boss that this particular candidate was the right person for the job. I tried to present a reason why I needed to hire this individual on several occasions but I was told no. I took the same proposal and argument to the consultant, and it became a good idea. I cannot understand why I had to go through this, but the final analysis is that I hired the person I wanted. It says that when you are patient, you will get what you want in the long run. This is what my father told me as a little girl. It was like a sermon to me because I felt my entire life had been one that I had to come second or third. It is not fair for anyone to go through what I have gone through in my life. I have always thought what it will be like for anyone to be put in my shoes. I regard myself as a strong individual and can handle any challenges that come my way, but it was like being in hell for quite a while for me. I had what it took to make my life and others pleasant, but I was in a constant battle.

The struggle continued as we moved into this new spanking building. The process of building a permanent facility was the next step. The hospital and board were now seriously making the necessary plans for the next and final phase. I felt everything was going on well, and coupled with the insurance policy we had on the hospital and its other properties, it was obvious to me that we had enough money to rebuild a permanent facility. The projected time for reopening of the new hospital was three years from the initial disaster. I kept myself informed with the process through several meetings which I felt excluded from intentionally. As time went by, the hospital was not making any money with its temporary facility. Some folks in the community felt the building was not sturdy enough to hold itself and did not feel safe in the building. The employees were so frustrated that they did not promote the building, which resulted in us not meeting the patient capacity of seventy-six, which we needed to stay profitable. We had also lost a lot of our specialty doctors due to the fact that the tornado affected them and resulted in them moving to other locations to make a living and feed their families. It was expected for us to face financial issues during the first few months, but the problem persisted. I first got the information about trying to sell the hospital from my boss when we took a trip to Albany, Georgia, to make a presentation about the hospital. He indicated it was only an option that he was thinking about. I did not understand. Why sell the hospital when we had insurance coverage as well as committed funding from FEMA and GEMA? I knew I did not have any input into the matter even though I was one of the team players, but it did not come to me as a surprise. I did not think anything about his discussion to me until a few weeks later, when it was announced in our regular monthly board meeting. I also found out that the other senior management personnel were in fact aware of the possible sale of the hospital. This was not a surprise to me because I was always kept in the dark and usually found out about the happenings of the hospital from the good old boys although they were in middle management position. Although I hated the fact that individuals that were working under my supervision had more power than myself, I had no choice but to rely on them for information.

It became another problem of not knowing how and when the hospital was going to be sold because it was the only job I knew and loved, and at my age and also due to the demographics, I knew it would be very difficult for me a find another job. I had gotten use to the environment as well as the people of the community. I had a connection with too many people and felt they depended on me not only for their health care, but they had become my family. I felt at home because they reminded me of my childhood surroundings. I could not figure out why we wanted to sell the hospital when we had enough money to build a new hospital. I came up with many scenarios of what I thought was going on since I could not get a clear answer from any of my peers or boss. All I could do at that moment was continue working as usual and let the chips fall where they will. I must also say I was afraid of losing my job because I was responsible for my family and could not imagine not providing for them. I had too much responsibility and started planning in my head what to do in the event I lost my job. On the other hand, I felt confident that if the idea of selling the hospital went through, I had enough contacts to help me secure another job. I felt I had done a tremendous job throughout my career, and although I felt uncertain about the future, I knew it would be a short time before someone or some health organization offer me a job. In the event the unthinkable happened and I lost my job, I had accomplished so much that if the new organization came in, they would immediately hire me. I felt I had been the face of the hospital in terms of involvement and contribution both to the community and to our employees. I guess I was more than confident in my performance and could not imagine what was ahead of me.

It became apparent that the hospital was going to be sold after all when senior management was asked to cut their staff. I did not have as many staff members to cut from my departments but was given an ultimatum to get rid some of my workers that had worked for me for a long period of time. It was one of the most difficult things I had to do because some of them were single parents and depended on their jobs for survival. Before I had the opportunity of discussing what was going on with my employees, word had already gone out about lay-offs at the hospital. It was a common practice for information to be leaked out beforehand because only certain people, regardless of their position, were kept informed of the happenings at the hospital. One would think that as an executive of the hospital, I would know what was going on. Most of the time, I only found out what was going to happen at the board meetings which I attended. To be honest, I felt the board did not care about the situation at the hospital and relied on management to make the decisions. For some odd reason, I only observed a few times when the board went against management in all of my twelve years as an executive at the hospital. I am not sure what their purpose was except to say they were members of a board that was the second or third highest employer in the county. It was also very prestigious to sit on one of the three hospital boards because Mrs. Lilian Carter, the mother of our thirty-ninth president who was a nurse, was one of the original board members. I thought the fact that the board never disagreed with management on some critical issues involving the health care presence in the county was due to the fact that it was a not-for-profit organization and they were not compensated for their services The only benefit they received was the occasional retreats that occurred once or twice a year at an exclusive location. This is all speculation on my part on the how and why the board did not have full control over management, but being there and dealing with the several boards for twelve years, I think I could be considered a fair judge of what I saw going on.

It became clear that the hospital was going to be sold, and no further attempts were being made to save it. As a result of this process, we lost a lot of our doctors and professional employees who immediately knew they had to find work elsewhere and not wait for the new company to clean house. Our census started to drop and the talk of the hospital that was going to acquire our hospital made it worse because they had a reputation of unfairness to their employees and wanted to basically buy every health care provider in their market area. They also had the reputation of control and had a regime that could be considered as those of some of the countries in the Middle East. This was not only a thought I had but one that was expressed by so many people. This particular hospital had been sued by a few doctors and made headline news which was broadcast on *60 Minutes*. Most folks at the hospital, after hearing the news that the hospital board was thinking seriously of selling to this hospital, did not think it was a good idea. As mentioned earlier, as much as the board gave the impression they were the decision maker, it was really the executive team at the hospital. This team at the hospital consisted of just a few who took matters in their hands and did not care what the community or employees thought about who to sell the hospital to. Although I was one of the executives on the team, I did not even sit in any of the planning sessions for the sale of the hospital. I knew of the final buyer of the hospital when the secretary asked me for my resume which was some of the documents necessary to close the deals—from all of the executive team. This was bothersome to know that I was part of the executive team and still could not be treated as one, even after I had made several complaints to anyone that would listen. My attitude moving forward was to not get mad at anyone but try to make the best out of the situation. It was not my purpose to change anyone or their opinions about how the world should be or how they should treat me for that matter. God was in control and saw everything that was taking place good or bad. My fate was in God's hands and crying or spending more energy in trying to read the minds of others was not the answer. Now that I knew some of the executive team were definitely going to lose their job eventually, I prayed that when the new buyers reviewed what I had been able to accomplished, they would realize I had accomplished a lot in my twelve years. I was also confident that I had enough support from the community, especially with our local and state senators, who I thought would make sure I kept my job. I had also become good friends with our thirty-ninth president and knew I could contact him or his wife for references if need be.

A few months went by with serious negotiations with the other hospital and our organization to make sure the deal was legal and accepted by both parties. The purchase was accepted by both parties, and the process of taking over began. It became apparent that the hospital, indeed, was going to be sold after being an institution in the community for over fifty years. I could not believe why this was happening and how the board could approve its sale when, in my opinion, we had better options; for example, build a smaller hospital with the help of FEMA and GEMA. We had approximately ninety million dollars in insurance money and help from the government to rebuild, so why make a deal with the devil? I was not part of the discussion and basically got my information from the other members of the executive team. I was hoping our board members would stand up for the people in the community and refuse the sale of the hospital, but they chose to listen to the management of the hospital and let the hospital be sold. I was not surprised by their decision because as long as I had been at the hospital, the board did not and never put the interest of the people of their community first. I felt that the board members came to meetings or sat on the board for one reason alone, which was get a good meal once a month or go on board retreats. These individuals were not paid for their services, due to the fact that the organization was not-for-profit, and they were not losing an income. In my humble opinion, no one seemed concerned about losing the hospital. We immediately went into the mode of selling the hospital and not trying to save it. The argument given was that we could not afford to build a new hospital even with the insurance money as well as help from other sources.

On or about March 13, 2009, more than two years after the destruction, the hospital authority unanimously adopted a forty-year lease agreement with another hospital, basically giving them the rights to purchase the hospital. This obviously was not final because the deal had to be approved by the attorney general with a ninety-day approval process. The deal required the purchasing hospital to commit twenty-five million dollars to the construction of the hospital. This agreement did not sit well with a lot of people in the community because the majority knew we had insurance to cover the cost of building and we had commitment from FEMA and GEMA to cover the balance of the cost of rebuilding. I, for one, could not understand the logic to getting a partner—as it was called—although I knew we were giving the hospital away. As the news got out in the community of the sale of the hospital and the dismissal of some of the executive team members, the press wanted to get information from the management team. It was difficult to get the CEO to commit to any interview so as vice president of public relations and marketing for the hospital, I made myself available to the press to answer to the best of my ability what I knew of the situation. Here is an excerpt of what I was asked by a local newspaper organization.

"The people of Sumter County voted for a hospital in the early 1950s and paid property tax bills to pay for this hospital—how did the people lose control over it? The hospital did not need any help with their $20 million expansion office space, etc., now you sold us out for 40 million with a net worth of about $119 million. Can you explain this?"

My answer to these questions were abrupt because I did not want to put myself in a position that would seem I was against the hospital. I found myself defending the hospital even when I knew were not working hard enough to save the hospital. The hospital had received enough funding from GEMA and other sources to continue with the construction of the new building, but for some unknown reason, certain people in administration and the board wanted to give the hospital away.

Now it was apparent that the hospital was going to be definitely sold after the ninety-day wait period required by the attorney general. One of the executive members on the team who was one of the negotiators of the deal left the organization and secured a job. At this point in time, I knew we had given up our hospital, but my gut told me I was going to be okay. I did not want to lose my job after working there for such a long time. This was the job I had worked the longest and actually felt I had made a difference not only for my family but for a lot of people. I did not want to start looking for work and start over again. I was up in age and felt it would be difficult to find a senior-level position in this area of the country. My youngest was still in high school, and due to the real estate crisis, I knew it would be difficult to sell my house. I had to look up to the Lord once again for direction on what to do. My anxiety level at this point in time was at an all-time high level. The thought of not getting up each morning and going to work although I hated doing so was unimaginable. I had to come up with a way of saving my job. I had done this for the past thirteen years and felt I wrote the book on convincing whomever to keep my job. I started making more phone calls to individuals I knew had the power to help me. I contacted the mayor, senators, councilmen—who's who in the community that I felt was on my side to remind them again that I felt my job was being eliminated and gave them my story on why I should keep my job.

A week later, I got an e-mail from my boss giving me a date when my services would no longer be needed. After I checked with the other vice presidents of their departure date, I realized that my day was two weeks earlier. I sent an e-mail to my boss asking to stay another two weeks, but he refused and told me the decision was final. I also at this point found out that I was given a six-month severance pay while others had up to nine months' severance. Is this justice? You judge for yourself.

On June 9, 2009, four days before my twenty-second wedding anniversary, with my walking papers in hand, I collected the rest of my belongings in my car and drove away from the hospital that I had been a part for thirteen years. It was one of the saddest day of my life next to the day I lost my beloved father in 2001. I could not imagine not getting up every day and making the fifty-mile journey to work, driving through the country road that I had done for so many years. It was more than losing a job because the people I worked with were like family to me. The patients needed me, and the young children in the community needed me, I thought. I could not wait to make a difference or talk to a young child about his or her future or be part of their success as I had done with some of the people that worked for me. I wanted to be like my father who opened so many schools in Africa and devoted his life to the church and education. I remember when, at sixty-eight years old, my father was forced to retire, and he took his employer to court because he felt he had more years to give and more people to educate. It was all about making someone else get better and pass on to others. It was not time for me to leave an environment that had treated me like a slave yet rewarding because I felt I made a difference in so many people's lives.

Why do we still have problems in relating to different races especially in the United States of America that slavery was supposedly abolished hundreds of years ago? Why would someone as powerful as individual make an outrageous statement about the president of the United States authenticity of his birth place? Would he criticize another president if they were of a different race? Is this person getting away with his outburst about the president's citizenship even though the State of Hawaii confirmed his live birth? Is it due to the fact that he or she can afford to pay the most expensive attorney to represent him? Why are we still accepting bad behavior from people who think because they are of a certain race they can say and do whatever they please? I can go on forever, quoting many incidences that have taken place in my lifetime and truthfully cannot understand why and all I can do is live with it and accept it as one's opinion. This powerful individual said on CNN that it was his opinion and everyone is entitled to their opinion. I remember when he at one time in the early nineties had lost most of his wealth and came to one of the poorest cities in the United States (Gary, Indiana) to buy a casino. My true opinion is that he only came to Gary to acquire wealth and did not have any care or concern for the people of Gary, Indiana. God is the only one that has answers to why people are treated one way or the other. Why are we still debating the race issue in the United States of America? Why has the justice department demanded an investigation on the police department of the city of New Orleans? Could it be because there is truly a problem with the way the Blacks are being treated? Are they going to be able to solve the problem, or is this investigation done just to say and investigation was conducted? Why not spend the funds to help the poor around the New Orleans area instead of spending the money on people who have enough already? What it is going to take to make sure that race relations is handled in the proper way in the United States of America? Why are we fighting battles in other parts of the world when we have our own battles to fight right here in the United States of America? In the Book of Matthew, Chapter 7, it says, "Stop judging, that you may not be judged. For as you judge, so will you be judged, and the measure with which you measure will be measured out to you." This verse begs the question of whether I am judging or have been judged. All I can say is I wrote this book from the experience I have lived. Nobody asked to be poor or Black, White, red, or indifferent. We are all God's children, if you believe there is a God. I wish I could say things will change, but this saga will continue, I know, in my lifetime. It has been three years since I lost my job. I cannot begin to explain why I am unemployed when I was one of the best at what I did. My peers from the hospital are all employed and have been for a while, and I am the only one still looking for employment. I can only say the reason I cannot find a job is because I am a Black woman in the South, and as much as I performed, I am still without a job. I have the masters in business administration, experience in health care administration, but yet no job offers. Could the reason for me not working be related to the economy, being a woman, Black, not liked by my previous employer? People might think because we have a Black president, Blacks would have a high level of employment. The only difference I have seen since our president took office is that they are more Black reporters or correspondents than usual. Is this a coincidence or might this be because we have a Black family in the White House? Whatever the case is, the bottom line is we, as humans, have to handle situations as we get them. We should not try to put blame on anyone for the trials and tribulations that come into our lives. I am sure there are people who are going to try to evaluate this accounting of my perception of what has happened to me. I am certain Dr. Drew, the television personality would have his take on my story; Anderson Cooper might have a few panelists to give their take on my story. This is all good with me because we all have different ways of expressing ourselves as well as dealing with different situations in our lives. These are all factors that are dealt with differently by each person. God is the only one that has answers. I pray for my children each day that maybe they will see a change in the future and be accepted as human beings regardless of their race or financial status. I would venture to say that they might not have the same experience. There is a saying in my culture that says, "We want our children to have more than we had." Have we, as parents, ever thought our children might be satisfied with what they have? Is having more education or money the answer to our problems? Are the rich happier than the poor? Is being Black better or worse than the other colors? Is my story worse than one that has been experienced by another person? In my situation, I can truly say that God gave me the grace to move on and make it this far. Only by the Grace of God was I given the opportunity to experience this live and have the privilege to meet and touch a lot of lives alone the way. In the final analysis, *still colored* is not so bad after all.

The final chapter of my book is dedicated to my father, Thomas Sona Ngu. My father was born around 1913 in Kumba, a farming town in the southwest of the United Republic of Cameroon. He had just one brother. At an early age, his father died and left him to be raised by his mother and uncle.

At an early age, his mother singlehandedly made sure he went to school instead of working on the family-owned farm. Not a lot of people attended school because it was not what was important, and male children had to stay home and work manual labor to support their parents and the community. Due to the fact that my grand-mother was determined to raise her two boys with an education, she made sure that he received an education from the government. After finishing his primary education, he moved to another city and decided to join the Basil Mission Church which is now known as the Presbyterian Church to further his education in teaching. Most of his friends decided to go into politics, but his vision was in education.

In the early 1930s, my father married his wife, Esther Mafor, and continued his education as a teacher with the Basil Mission. He received his certificate in higher education in a town called Keke. He was one of the youngest students in his class and went on to work as a teacher for the Basil Mission. His primary duties in the beginning was to go around the western cities of Cameroon and open schools and train teachers. He did this particular job for years and with his growing family. He told me that he had to travel to many rural towns by foot, and on several occasions, the roads were so bad it took days and weeks to move from one location to the other.

Thank God my parents used the Scriptures to raise all of us thirteen siblings and three adopted children, which personally enabled me to excel in life. My father instilled in me that the myth of Christianity being a White man's culture was totally of no effect in our world. No inferiority complex was registered all along my upbringing till date. It was instilled in us as children as written in 2 Corinthians 6:14, "Do not be mismatched with unbelievers, for what partnership is there between righteousness and lawlessness? Or what fellowship does light have with darkness?"

It was a known fact that God was the head of our household. My parents opened a church in the town that we lived in, and it was mandatory that all children participated in as many church activities as possible. We all sung in the church choir as soon as we could talk, and there was no exception.

We lived in a polygamous society in Cameroon, but we were not exposed to that type of lifestyle because my father said we were Christian and that type of lifestyle was for the unbelievers. It is hard to believe that my father chose to not participate in polygamy although the majority of the people practiced polygamy. I was privileged to have parents that were blessed, so-to-speak, to be fully equipped with the Word. As with most African cultures, they believe was that male children were more accepted than female children. In my father's case, he regarded all his children as equal. As a female child, in the ninth position with three senior sisters and three younger siblings and seven brothers, we were all given the same opportunities and encouragement to do what we wanted.

In my culture, a female child, in those days, was considered a man's property. In other words, the female child was brought up to depend on a man and did not deserve much academically. My dad, may God bless his soul, gave us the same treatment he gave to the male children. My father's motto was that every child, male or female, would be trained to be financially independent. All of his daughters went through higher education, three attended American universities, one in a Nigerian university, and one in Nursing and Midwifery University in Cameroon. His thinking was that all his daughters should be educated so that they are independent and should not be placed behind the man as the adage goes, "behind every successful man is a woman."

I do have volumes of what I received from my dad, but I will share with you as far as I can remember, and hopefully, you will be amazed of who he was and what lessons you can derive from this great man. Everything he did was based on biblical norms. He told me, "An orange tree produces orange fruits, and consequently, orange juice."

Furthermore, he always insisted on the value of education. He made sure I knew that knowledge was gained from continues learning. He let me believe that acquiring knowledge involves cognitive processes, communication, and logic. An example is when I graduated with a master's degree in business administration, it felt to me like I had conquered the world, but the question my father asked me was, "When are you going to start on your PhD?"

I immediately told him that very few people have an MBA, and his response again was, "You never stop learning."

He also let me know that knowledge is the understanding and awareness of something. It refers to the information, facts, skills, and wisdom acquired through learning and experiences in life, and it has no end. I was very inquisitive and bold to ask my dad questions that my older siblings could not ask. In one instance, he told me I need to listen more and be patient. He incorporated the Word of God in every conversation he had. He also let me believe that the Word of God was the most valuable knowledge a human being can possess but also know that simply being aware of God's existence is not sufficient; it must encompass the deep appreciation for and a relationship with him. These important values reminded with me and encouraged me to become a Stephens Minister at my church. Going to church services each Sunday gave me an opportunity to hear my father preach the sermon and make the church announcement. I also listened to the choir group my mother started. My father and mother made all the children learned a few songs like "Jesus Loves me this I know, for the Bible tells me so," and "Stand Up for Jesus." We all had to memorize these songs or else we will get in serious trouble.

My father and my mother amazingly sheltered their children and made them their priority. I, in turn, use the same theory and practices my parents used to raise my three children. Education and respect was the norm my husband and I relied on and made sure our children attended church services each week. We both ushered in the church and taught Sunday School classes. I remembered my husband had to carry our youngest son on his back during the services to perform his usher duties because my son did not want to stay in the nursery.

My father, the man behind the scene, touched the lives of so many people in the English-speaking Cameroon. Most of the leaders of Cameroon were taught in school by my father. He was one of the founding fathers of the Presbyterian Church in Cameroon. The first prime minister and former speaker of the House and the vice president of Cameroon were all of his personal friends. My father received two medals of humanity from the president of Cameroon, but his greatest achievement was his God and children. Because of my father's relentless push, I was able to be the first African American to occupy the position of vice president at Sumter Regional Hospital in Americus, Georgia. I also had the opportunity to interact with the thirty-ninth president of the United States of America; to sit on Mrs. Roslyn Carter's Board for Caregiving; to negotiate with Miller Fuller, founder of Habitat of Humanity to establish an office in Cameroon; and to work with Congressman Sandford Bishop on numerous projects to improve healthcare in rural Georgia.

In conclusion, my father gave hope to countless people in Cameroon and never asked for anything but a thank you. He made me the strong individual I am, and all my achievements in life were due to my father and mother. He told me that you can be anything you want if you believe in God and work hard.

Comfort and Granddaughter Aubrie

Comfort, Amie and Aubrie

Imani Ngu, my parents youngest Grandchild

Spurgeon Green IV at a modeling competition

My Son's Law School Graduation

My Son's Law School Graduation

My Son's Law School Graduation

My Wedding Day

Rev. Dr. Robert Ngu walking me down the aisle

My Wedding Day

My Wedding Day

My Wedding Day

My young children

My young children

My family on vacation

My family on vacation

My Daughter and her Grandparents

Comfort and Margaret at our nephew's wedding

Comfort, Margaret and Elizabeth

Comfort's brothers Jospeh, Dr. Lawrence Ngu

Comfort graduating with and MBA

Comfort's parents Thomas Sona and Esther Mafor

Thomas Sona, Esther and their youngest child

A memorial of all of Comfort's siblings

Visiting Cameroon

Bamenda, Cameroon

Bamenda, Cameroon

Comfort and John Edwards, Presidential Canidate

www.americustimesrecorder.com

Volume 129, Number 50

JETS
Both teams move on to Semi-finals
PAGE 1B

VICTOR GREEN
Supports SRH Souvenir Brick Drive
BELOW THE FOLD

GOLDEN PEANUT AUCTION
Take home a piece of history
PAGE 3A

Americus Times-Recorder

SERVING SINCE 1879

WEDNESDAY, FEBRUARY 27, 2008

'PATH OF FURY'

The *Americus Times-Recorder* has sold out of its pictorial book on the March 2007 tornado, "Path of Fury." The Americus Welcome Center, located in the Windsor Hotel, downtown, still has some copies of the book for sale.

INSIDE TODAY

Day two of Neal trial gets underway

Closing remarks, verdict expected today

By GENIE COLLINS
genie.collins@gaflnews.com
americustimesrecorder.com

AMERICUS — Day two of the Shavis Neal trial got underway

Tuesday in the large courtroom of the Sumter County Courthouse.

Neal is charged with murder, felony murder, four counts aggravated assault and possession of a firearm during the commission of a crime, in connection with the 2006 shooting and murder of Dekeldric Thornton at the Americus Elk's Lodge.

Southwestern Judicial Circuit Superior Court Judge George Peagler presided over the trial with District Attorney Cecilia Cooper prose-

cuting and Ron Beckstrom, a criminal attorney from Albany, acting as defending counsel for Neal. Cooper was assisted in the case by District Attorney's Investigator Calvin Mansfield.

The first to testify Tuesday was Americus Police Chief James Green. He testified that although he was originally not on duty the night of the shooting, he was called by dispatch and Maj. Richard McCorkle, another investigator with the Americus Police Department.

Green testified that when he came to the Elk's Club, there was a crowd, but nothing really caught his attention. Then, he went to the emergency room at Sumter Regional Hospital, where he witnessed "yelling."

While he was at the hospital speaking with various individuals about the shooting, Green testified that Commander Nelson Brown left the hospital, and the chief said

Please see NEAL on page 5A

BLACK HISTORY COLLECTION: TEMPLE UNIVERSITY

Black History, page 7A

Former NFL star, Americus native shows his support of SRH Souvenir Brick Drive

From Staff Reports
americustimesrecorder.com

AMERICUS — Former NFL player and Americus native Victor Green recently pledged his support of to Sumter Regional Hospital's (SRH) Souvenir Brick Drive. Green played college football at the University of Akron. He entered the NFL in 1993. By 1994, Green was the Jets starting strong safety. He played for the Jets through the 2001 season. Green played the 2002 sea-

Submitted Photo
Here, SRH VP of External Operations/Marketing Comfort Green presents a souvenir brick to Green.

son with the New England Patriots and the 2003 season with the New Orleans

Saints. On April 4, 2006 he again signed with the Jets so that he could retire on the

team he started his career on the

Please see GREEN on page 6A

Antique Show-Sale is March 28-30

From Staff Reports
americustimesrecorder.com

LESLIE — The 20th annual Dogwood Antiques Show and Sale in Leslie is March 28-30.

The event will be held from 10 a.m.-6 p.m. Friday and Saturday, and from noon-4 p.m. Sunday at the Leslie Civic Center, said Joely Ledger, coordinator of the event.

Each year, the event is sponsored by the Town & Country Garden Club.

"We're expecting a good year," Ledger said, in a phone interview Tuesday. "We want to invite the public to come."

She said this year's show features approximately 20 dealers, but there will be no appraiser on hand to

ARTS COUNCIL SUPPORTS SUMTER SCHOOLS

Education, page 7B

Recovery events planned for weekend

Citizens have opportunity to remember, reflect on tornado

By BETH ALSTON
beth.alston@gaflnews.com
americustimesrecorder.com

the past year.

From 5:30-7 p.m. Friday, at Georgia Southwestern State University's Storm Dome, Gov. Sonny Perdue is the guest speaker and will present a proclamation.

The event — "We Remember: A Recovery of Recognition and Recov-

state Sen. George Hooks, D-Americus.

Also on the program are Americus Mayor Barry Blount and Sumter County Board of Commissioners Chairman Bill Bowen, who will give the welcome, followed by the presentation of the colors by an Honor Guard representing local public safety agencies. The invocation will be given by the Rev. Norris

Comfort and Victor Green

116

Comfort and Congressman Sandfor Bishop

SRH's new marketing chief has interesting background

By STEPHEN GURR
Staff Writer

If you have trouble placing Comfort Green's distinctive accent, there's a good explanation.

Sumter Regional Hospital's new director of marketing and public relations comes to Americus from Illinois – by way of Minnesota – by way of England – by way of her birthplace, Cameroon, Africa.

Mrs. Green, 39, grew up the daughter of teachers at an English-speaking Presbyterian school in the West African country, which borders Nigeria and is about the size of Nevada. At age 13 she was sent to a grade school in England, where she lived with an Irish couple and stayed through high school. She opted to move to the United States after graduation, in part for its broader selection of colleges, and ended up choosing Mankota State, about 90 miles south of Minneapolis.

After graduation she worked as an insurance adjuster and met her husband, attorney Spurgeon Green III. She later earned a master's of business administration in marketing and together they moved to Joliet, Ill., where Mrs. Green worked as a marketing director for Complete Health Care HMO from 1987-1994.

Mrs. Green's in-laws moved to Perry six years ago, and she and her husband followed a few years later. While the couple and their three children live in Perry now, they plan to find a home in Americus soon. She began working for the hospital in September.

Mrs. Green says Perry has been a good place for children Amy, 9, Spurgeon, 7, and Thomas, 5.

"It's probably a better community to raise children," says Mrs. Green of South Georgia. "It's rural. It's quieter. It's smaller. It's not as fast-paced [as Joliet]."

Two years ago she took the children back to Cameroon, where her parents still live.

"My son thought it was too hot," she says. "It was a good experience for them to see a lot of the more underprivileged children there. But they were ready to get back to the states."

A former soccer forward and basketball player, Mrs. Green enjoys coaching eight-and-under soccer and playing hoops with her kids. She says education is a big concern for her.

"That's my mission in life, to make sure that young kids go to school," Mrs. Green says. "A high school degree is not enough – you have to stay in school. It's very critical, and anybody can do it. You have to shoot for the stars."

Mrs. Green exudes excitement for her new employer, its involvement in the community and the monumental expansion of facilities.

"This is a great hospital," says Mrs. Green, who has worked for in the field for over a decade. "We're really working for the community – and not just Sumter County. This hospital features extremely qualified physicians and staff, and we want people to keep that in the back of their minds when they come here."

And it's her job to make sure they do.

Times-Recorder/STEPHEN GURR

Comfort Green, Sumter Regional Hospital's new director of Marketing and Public Relations.

SRH hires Marketing Chief

business administration - marketing and Dallas Cody Timmerman, bachelor's of business administration - marketing.

Selection Committee named for SGTC GOAL program

From Staff Reports
americustimesrecorder.com

AMERICUS – Cynthia Carter, coordinator for the Georgia Occupational Award of Leadership (GOAL) program at South Georgia Technical College (SGTC), has announced the members of the screening and selection committees that will interview students nominated for the college's annual GOAL award.

Named to serve on the selection committee are Don Smith, assistant to the president, SGTC; Comfort Green, vice president of External Operations & Marketing, Sumter Regional Hospital; Angela Westra, Family Connection coordinator, Visions for Sumter; Kim Dunn, vice president/branch manager Forsyth Street, Sumter Bank & Trust; and Lorenzo Waters, CIS manager, Glover Foods.

The screening committee consisted of Lemond Hall, director of Evening Operations; Deborah Jones, Distance Education coordinator; Sandhya Muljibhai, Workforce Investment Act coordinator; Mark Brooks, director of Administrative Services; and Diane Trueblood, Media specialist/Crisp County Campus.

The objective of the GOAL program is to recognize and reward excellence among the almost 150,000 students attending the colleges of the Technical College System of Georgia. GOAL, which is now in its 37th year, was the first statewide program in the nation to honor outstanding students in technical education.

"To be considered for the GOAL award, a student must be nominated by his or her instructor," explained Carter. "The purpose of the screening committee is to evaluate the nominated students to determine who qualifies to move to the next level of the competition."

In making their choices, the screening committee members consider each student's grades, attendance and class performance along with character, leadership potential, personal goals and enthusiasm for technical education.

The students who make it through the screening committee will then be interviewed and evaluated by another GOAL committee – the selection committee – which is comprised of leaders from local business and industry. The selection committee chooses one winner from SGTC to compete at the statewide level with students from the 32 other state technical colleges and four Board of Regents colleges with technical education divisions.

"All of these students are exceptional individuals or they would not have been nominated," said committee member Don Smith. "The GOAL program is an excellent means of recognizing the outstanding students enrolled in the state's technical colleges, and I appreciate the opportunity to be involved in the selection process."

In May, the SGTC GOAL winner will take part in the technical college system's annual statewide GOAL competition in Atlanta. At that time, one student will be named as the state's GOAL winner and earn the recognition as the system's 2008 student of the year.

The state winner will also take home the grand prize of a new vehicle from Chevrolet, the statewide corporate sponsor for the GOAL program.

Members of the South Georgia Georgia Technical College GOAL Selection Committee are, from left, Don Smith, Comfort Green, Angela Westra, Kim Dunn and Lorenzo Waters. *Submitted photo*

South Georgia Technical College committee

Comfort's work badge

Five years of service at Sumter Regional Hospital

Comfort being interviewed

Service at Sumter Regional Hospital

Halloween celebration at SRH

Breast Cancer Awareness at SRH

SRH Fundraiser

Comfort and Leonard Pope, NFL player

Comfort and Leonard Pope, NFL player

Comfort and family with Victor Green, NFL player

Comfort and Bruce Johnson

Milard Fuller, founder of Habitat for Humanity

Sumter Regional Executives

Judge Peagler at SRH

Attorney General Thurbert Baker, Comfort and Spurgeon Green III

Attorney General Griffin Bell

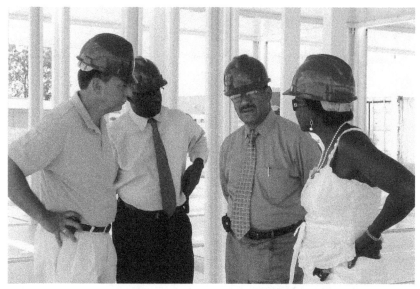

Congressman Sandford Bishop and the SRH executives

Comfort and Congressman Sanford Bishop

Comfort and Presidential Candidate John Edwards

President George Bush visits SRH

Spurgeon Green III, and Sam Donaldson, ABC News

Comfort, Spurgeon III and Former Macon, GA Mayor C. Jack Ellis

Comfort, Author Rudy Hayes and Sam Donaldson, ABC news

Comfort, Spurgeon III and Spencer at Presdient
Carter's 75th birthday party

The
Carter Celebration Committee
requests the pleasure of your company
at a reception honoring
President Jimmy Carter
"A Taste of the Carter Years"
Friday, the first of October following the Birthday Gala
The Windsor Hotel
Americus, Georgia
Black Tie Optional

For security reasons, please present this card for admission
Admit only after ten-thirty

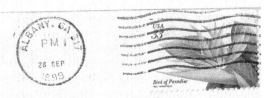

Mrs. Comfort Green
Public Relations
Sumter Regional Hosp
Wheatly Dr.
Americus, Ga. 31709

President Carter's 75th Birthday Party Invitation

Comfort being interviewed

Comfort and Kenneth Cuts, Sanford Bishop's Press Secretary

Judge Peeler at a Bird in Hand fundraiser

Comfort and Elizabeth Edwards

Comfort and family visiting the Carter's church in Plains, GA

The Carters visiting SRH after the destruction

Comfort and First Lady Rosalynn Carter

Comfort and First Lady Rosalynn Carter

Comfort and First Lady Rosalynn Carter

Comfort and First Lady Rosalynn Carter

Comfort and Anne Spear, Hospice director

Comfort and Willie Faust my Marketing Director

Comfort and Dr. Stephanie Brown

Sumter Regional Hospital Marketing Team

Comfort's assistants Brandy and Chasity

Comfort, the CEO, and the VP of Nursing at SRH

Reverend and Mrs. Faust

Comfort in her office

Comfort as Grand Marshall in Americus

Comfort at the hospital Interm

Comfort at the hospital dedication

Comfort and Spurgeon at SRH

Sumter Regional Hosptial after the destruction

Destruction of SRH

SRH Executives

Destruction of SRH

Destruction of SRH

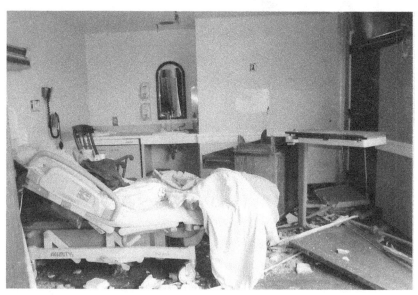

Destruction of SRH

Destruction of SRH

Destruction of SRH

Destruction of SRH

Destruction of SRH

Destruction of SRH

Destruction of SRH

Destruction of SRH

SRH Interim Hospital

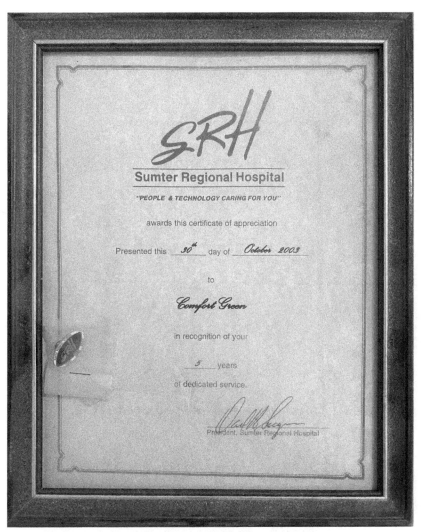

Five years of Service award from SRH

Executive Healthcare Award

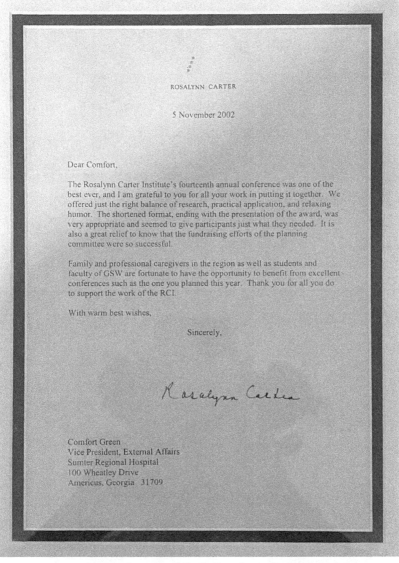

ROSALYNN CARTER

5 November 2002

Dear Comfort,

The Rosalynn Carter Institute's fourteenth annual conference was one of the best ever, and I am grateful to you for all your work in putting it together. We offered just the right balance of research, practical application, and relaxing humor. The shortened format, ending with the presentation of the award, was very appropriate and seemed to give participants just what they needed. It is also a great relief to know that the fundraising efforts of the planning committee were so successful.

Family and professional caregivers in the region as well as students and faculty of GSW are fortunate to have the opportunity to benefit from excellent conferences such as the one you planned this year. Thank you for all you do to support the work of the RCI.

With warm best wishes,

Sincerely,

Rosalynn Carter

Comfort Green
Vice President, External Affairs
Sumter Regional Hospital
100 Wheatley Drive
Americus, Georgia 31709

Letter from First Lady Rosalynn Carter

Comfort's Bachelor's Degree

Comfort's Master of Business Degree

Comfort and Evelyn Mambo Ekobena, friends for 45 years

About the Author

Comfort Green was born in Cameroon, West Africa, to Thomas Sona and Esther Mafor Ngu. She was the ninth of thirteen children. Her father was a teacher by profession, and her mother, a homemaker. Comfort was raised by two independent parents that instilled in the children that God came first and education was equally as important.

At an early age, she was sent to England to complete her primary and high school education. At eighteen years of age, she came to the united States to begin college at Worthington Community College in Minnesota. She received her associate of arts degree a year later and transferred to a university in Mankato, Minnesota, where she received a bachelor's degree in business administration, and subsequently, a master's degree in business administration. She also took several courses in political science in preparation for law school.

After spending ten years in college, she secured a job with State Farm Insurance in Madison, Wisconsin, as a claims adjuster. After a year with State Farm, she moved to Joliet and got married to her husband of thirty-five years presently and took a job with his father as a medical recruiter for his health management organization. At this point in her career, she decided to concentrate in the healthcare spectrum.

In 1995, her husband and three young children moved to Perry, Georgia, because they wanted to raise their children in a smaller and safer community. Comfort acquired a job with Flint River Community Hospital in Montezuma, Georgia, as a practice manager for two years.

In 1997, she moved to Americus, Georgia, about ten miles from the home of our thirty-ninth president, Jimmy Carter, where in ten months, she was promoted to vice president of external operations.

Comfort is a mother of three adult children and has one granddaughter.

CPSIA information can be obtained
at www.ICGtesting.com
Printed in the USA
LVHW011955140723
752117LV00004B/600